SNOW

SNOW
Ruth Kirk

University of Washington Press

SEATTLE AND LONDON

For Saboro Yamamoto,
who served at the meteorological station
atop Japan's 12,388-foot Mount Fuji,
highest in the world to be manned year round;
and for Harumi, his widow, and Yutaka, their son

Originally published by William Morrow and Company, Inc., 1977
First University of Washington Press edition,
with a new Preface by the author, 1998

Library of Congress Cataloging-in-Publication Data
Kirk, Ruth.
Snow / Ruth Kirk.
p. cm.
Includes bibliographical references and index.
ISBN 0–295–97734–5
1. Snow. I. Title.
QC926.32.K57   1998   98–3475
551.57'84—dc21   CIP

The paper used in this publication is acid-free and recycled from
10 percent post-consumer and at least 50 percent pre-consumer waste.
It meets the minimum requirements of American National Standard
for Information Sciences—Permanence of Paper for Printed
Library Materials, ANSI Z39.48-1984. ∞

Photographs in front matter by Ruth Kirk

PREFACE

READYING *SNOW* FOR REISSUE TWENTY YEARS AFTER ITS
original publication has stirred memories. When the book first
appeared in 1977, New York City was undergoing the white
siege of a blizzard and the *Wall Street Journal* ran a review of
the book on page one.

For the next several weeks the pattern of requests for radio
interviews let me track the path of storms accosting the North
American continent. Nobody seemed pleased with their current fate; interviewers' and listeners' interest in hearing about
snow's remarkable attributes was secondary to their talking
about its all too real inconvenience. Yet snowfall brings blessings. It acts as a blanket for overwintering plants and animals,
a highway for some people, a playground for others; an aesthetic
grace note, a water source, and a climate engine (because of its
reflectance, which influences atmospheric circulation processes).

At the outset of working on the book during the early 1970s, I realized that looking under "Snow" in the card catalogues and journal indices of libraries was not the right research approach. Rather, the relevant literature lay scattered among headings as diverse as Cloud Physics, Anthropology, Cryobiology, Arctic and Antarctic Exploration, History of Skiing, and so on.

The information also was held in the minds of people who had experienced snow in remarkable ways. I remember meeting Sir Charles Wright, at the time one of only two living survivors of the 1911–12 Scott expedition to the South Pole. Two things he mentioned stay particularly vivid in my mind: nightly, the men had to thaw their way inch by inch into their frozen sleeping bags; and when they glanced at the sky on stepping out of the tent in the morning, their clothing instantly froze and their heads were locked in a looking-up position that was torture for a man hauling a heavy sledge.

❄

The original book is reprinted here "as is" rather than with revisions, which would have raised the retail price disproportionately to the importance of the changes. Some individual photographs and captions are new, but four short photo sections remain. In the twenty years that have passed, word usage on several of these pages has grown obsolete. Identifying them is a measure of time's passage. "Man" is no longer used as a synonym for "people" or "humans." Eskimos prefer to be called Inuit, their own name for themselves; Lapps prefer their own name, Sami. Reference to the Soviet Union is out of date. The geodesic dome at the South Pole is no longer "new." Tectonic plates are accepted as reality rather than as the "recent theory of continental drift." Cloud seeding, which began in the 1940s as a way to increase precipitation, remains experimental but has fallen out of favor overall. Understanding what causes ice ages has grown more complex rather than clearer with increasing knowledge. Researchers at the California Institute of Technology recently found magnetic evidence of at least two episodes of glaciation so severe that the equator partially froze—one episode about 2.2 billion years ago, the other less than one billion years ago. Possible reasons for these ancient super-chills have so far escaped detection.

During my research for the book, I was particularly delighted to find that monkeys make snowballs. In the wild, young macaques have been observed rolling snowballs and carrying them around. In captivity, macaques at the Primate Research Center near Portland, Oregon, have made large snowballs, then sat on them. Now I have found that animal keepers at the progressive Arizona–Sonora Desert Museum near Tucson bring snow from Mount Lemon for the sole purpose of letting the animals in their charge play with it. They report varied reactions. "Treats" (food) buried in the snow by the keepers serve as effective lures, encouraging animals to investigate the unfamiliar. Buck deer spar with snow mounds presented to them. Otters jump on top of three-foot snowmen, playing, whether or not carrots are hidden within. Bears are leery; they can spend more than an hour checking and contemplating, then eat the snow and play with it. Jays peck and pull out strands of grass. Vultures "have lots of fun." Small cats seem indifferent. Pumas show interest in the hidden food, but not in the snow itself.

For human reactions to snow, I originally had hoped to find people somewhere with adaptations to snow that would match the sophistication demonstrated around the Arctic rim by the Inuits, but of course I failed. The Alacaluf and Ona people of Tierra del Fuego reacted to snow mostly by stoic acceptance of it, more resignation than adaptation; and Inuit people themselves now put aside many of their traditional ways of dealing with snow. Kids go out sledding with no apparent lower-body clothing other than jeans, and men set forth on winter snowmobile expeditions expecting to return by nightfall and therefore taking inadequate provisions for breakdowns or other delays.

In the 1970s, I talked about winter clothing and snow recreation with Jim Whittaker, then manager of Recreational Equipment Inc., in Seattle. The company had one outlet, indeed had just moved from its upstairs, Fifth and Pike location in the downtown core to a somewhat outlying, fullsize, street-level store. As this century ends, there are forty-nine REI stores nationwide and a co-op membership of 1.4 million. In the late

1990s, I spoke with Michael Collins, REI's public affairs manager, about what is new in cold-weather gear. Twenty years ago, Gore-Tex was a new concept in waterproof, breathable outerwear. Now it and its counterparts are well accepted. Polypropylene no longer is the only manufactured fiber to replace wool and silk for long underwear. And the wool knickers and shirts that Whittaker described as regaining popularity in the 1970s have been replaced by synthetic fleeces, which Collins says offer "the warmth of wool but are lighter, repel water better, don't stink when wet, dry quicker, and feel better against the skin."

Snowboards were new when Jim Whittaker described them to me; Michael Collins says they are now "well engineered," with an ability to edge into snow as skis do, and they have new bindings that provide better control. A snowboard team even took part in the 1998 Winter Olympics. Today's "hottest category" label belongs to snowshoes. What was "old" has become "new": aboriginal peoples across North America developed web snowshoes; across Eurasia, they developed rigid, wood skis.

On the serious side of changes within the last two decades is global warming, now accepted by most scientists as real, a result of increasing carbon dioxide in the atmosphere. If the winter freezing level in the mountains of western North America rises by, say, three degrees Celsius (a warming some experts expect in another thirty to forty years), watershed areas blanketed by snow are likely to be cut almost in half. What then will supply the growing population's need for municipal and irrigation water and hydroelectricity? Snowpacks may gain the belated respect and affection now accorded old-growth forests and wild salmon runs.

Lacey, Washington
May 1998

CONTENTS

CHAPTER

1

THE ROLE
OF SNOW

TRY TO DESCRIBE SNOW, AND IMMEDIATELY THERE ARISES A question of context. Shall it be the snow that falls as a veil and gently closes a household in upon itself, or the snow that streaks slantwise past the window and blows along the ground, blurring the surface and obscuring even the most familiar landmarks? The snow that to hydrologists is the near-perfect water storage system, or the sort that gives small boys ammunition for snowball fights? The snow used by Eskimos (and also by the U.S. Army in Greenland) as a building material, or that which becomes the headache of highway maintenance men?

Patterns of snowfall can't always be depended upon. On the night of December 30, 1976, Seattle skiers met to burn skis and chant a supplication for snow; yet that entire winter so little fell on nearby slopes that by the end of January a

Puget Sound yacht outlet was advertising: "Skiers, surrender. There's no snow. Your skis accepted as trade on a
sailboat." By Valentine's Day, 1977, the Washington legislature was considering cloud seeding as a means of whitening
the Cascade mountains and averting summer drought, and
Idaho politicians were protesting that the clouds, left alone,
were headed inland and whatever snow they carried was
theirs. Meanwhile, kitty-cornered across the country, the
Miami News issued a souvenir edition with a headline three
inches high proclaiming snow, and the manager of a Florida
nudist camp was quoted as commenting: "We're all huddled together and shivering a lot."

Northward, snow halted winter traffic. The city of Buffalo welcomed a twenty-man "Snow Liberation Crew" from
New York City's Sanitation Department, flown in with snowblowers and a scoop loader aboard a U.S. Air Force C-5A
transport. Drifts twenty feet deep buried cars and, even
where the snow was less deep, nothing moved. Businesses
of all kinds slowed, then stopped, prostitution included.
"We work only in cars," one hooker complained, "and all
the streets are blocked." Thus erratically did North America's winter of 1976-1977 enter the record books. Eastern
hookers and western skiers were stymied equally by the
vagaries of snow, whether an excess of it or a lack. Reindeer
at the Bronx Zoo escaped simply by walking up snowdrifts
mounded against their fence; yet in Waseca, Minnesota,
men had to attach wheels to sleighs so that the annual
sleigh and cutter parade could make its way down the main
street of town. Abroad, the Cairngorm Mountains of Scotland lay solidly blanketed above the twenty-five-hundred-
foot-level, the most snow in fifty years, and airline pilots
flying into Tokyo called passengers' attention to the oddly
white landscape below, with not just the mountains snowy,
but the lowlands and coast, as well.

Snow can be valued highly. During the Tokugawa era
of Japan, which corresponds to the European medieval period, the powerful daimyo Akimoto regularly curried favor
with the shogun by sending snow from Mount Fuji to the
court at Edo (the old name for Tokyo). Each summer his

horses and porters started from the high slopes heavily laden, and each summer they arrived at the end of the eighty-mile journey with little more than snowballs left. Nonetheless, there was enough to ice drinks and to shape into snow cones over which syrups could be poured, welcome delicacies on hot lowland days and a distinctive touch for the ruling court of the day. In a far more prosaic way, Japanese farmers even now value snow as a way to "read" how far along the season is. Those living in the mountainous region of central Honshu, northwest of Tokyo, set out their rice seedlings when melting snow on the flank of a certain peak takes on the shape of a man holding a hoe. In another area farmers wait for a black mare formed by a snow-free scree slope to stand beside a lingering snowfield shaped like a newborn colt. In the Innster district of the European Alps farmers similarly watch for a snow pattern called the White Scythe to appear on the slopes above their fields in late June. Then they mow their hay.

In Great Britain archaeologists have identified sites worthy of excavation by photographing from the air right after a light fall of fresh snow has accentuated irregularities of the earth's surface. "The effect is extremely transient, perhaps lasting only a few hours," according to Professor J. K. St. Joseph, Director of Aerial Photographic Studies at Cambridge University. "Sites show up in various ways: for example, snow may lie longest unmelted on the lee sides of banks and in pockets or holes amongst earthworks, so emphasizing features in relief. Again, with a buried site that has no trace whatsoever visible on the surface, ground that has been disturbed even thousands of years ago never returns to its original geologic compactness and it may have thermal properties that differ from surrounding soil, such as conductivity and specific heat. These characteristics may induce differential melting and make the site show up under a covering of snow."

Military men have a long history of using snow to their advantage, or of having battles inadvertently affected by its presence. Snow in the mountain passes stopped Alexander from continuing east into India in 330 B.C., and a century earlier Hannibal with his elephants incurred losses in the Alps because of the "white enemy." In the thirteenth century, snow blocked the Moors from Spain in their effort to enter France. In the fifteenth century, troops crossing St. Gotthard Pass to aid the Duke of Uri were "wretchedly devoured by an avalanche," and four hundred years later Napoleon crossing the passes en route to the Battle of Marengo instructed his advance guard not to "cry or call out for fear of causing a fall of avalanches." During World War I, Austrian and Italian forces purposely released avalanches in the southern Tyrol, once killing thousands of enemy soldiers in a single day. At the time censorship withheld news as to what sort of "heavy weapon" lay behind the incredibly devastating mortality of the slopes, but estimates of the toll during the three winters of the war now number sixty thousand, and some authorities say that figure is too low.

Among World War II situations involving snow perhaps the greatest long-range effect came from the Finns' near-defeat of the Russians in the three-month "winter war" of 1939–40. Observing it, the Nazis realized that although Russia ultimately forced its will onto Finland, the war nonetheless very nearly had been a case of David downing Goliath. From this they assumed the Russians to be weak, a grave miscalculation. On the contrary, the bitter Finnish experience taught the Kremlin that coping with snow must be fitted into army training and tactics, and by the time Hitler invaded the Russian homeland in 1943, Stalin's troops were as prepared as the Finns themselves had been earlier.

Nikita Khrushchev, looking back, remarked that in entering Finland he had believed there was no need to more than "raise our voices a bit and the Finns would put up their hands and surrender." In actuality one and a half million Russian troops were needed to achieve that goal, and of this huge number sent into Finland fewer than one-third lived to return home. The rest were killed, although the war lasted only 105 days and was fought in an exceedingly small land. "We won only enough ground to bury our dead," is how one Russian general summed up the whole campaign.

The Russians expected to sweep through Finland to Lapland in ten or twelve days; they already had annexed Estonia, Latvia, and Lithuania. In Finland shooting began on November 16, 1939, with the Finns outnumbered by a staggering ratio of forty-two to one and with little but pride and knowledge of the country—and of snow—on their side. Donning white camouflage coveralls sewn from bed sheets, Finns raced through the forest and over the fields on home-made cross-country skis to attack the Russians. Crawling on their bellies to shoot, they dragged their skis by thongs attached to their belts. They fought the enemy from the sides and the rear, and dove into snowbanks for concealment when they were spotted. They struck in the midst of blizzards and fogs and on moonless nights. The Russians fought only along roads and in open fields with tanks. Snow slowed their advance to no more than five miles an hour at best, and whenever they tried to cross frozen lakes they found themselves booby-trapped. The Finns cut openings through the ice to stop tanks, and on top of any that froze shut they laid strips of cellophane to fool enemy pilots into reporting the routes as impassable. From the air, the cellophane glistened like open water.

The Russians battled with one regiment lined up behind another and sought engagement with the enemy only according to war-college strategy. About their only innovation was to mount large steel boxes onto runners as troop carriers slid over the snow by advancing tanks. But Finnish bullets ripped through the steel and put an end to that one try at meeting snow on its own terms. Russian commanders went back to struggling conventionally with their heavy vehicles. To pack snow for them, they sent twenty or thirty men ahead on skis, and behind them another two hundred men trampling the snow with their boots so that trucks and tanks could get through. Or the men would shovel. Either way the battle against the snow angered troops and commanders alike and made them all the more vulnerable to the sudden raids of the Finns. In the end, the raids failed. March 1940 brought cease-fire, and Russia helped herself to what she wanted while Finland's brave troops skied home from the battlefields. The war was over, its duration and near-defeat a humiliation for the Russians. "The Finns have only their skis left," Stalin is reported to have said as the weeks dragged past. "Their supply of those seems never to run out." In fact, the skis very nearly had been enough to bring victory.

Across Siberia, Alaska, arctic Canada, and Scandinavia, people living close to the land invariably regard snow as an ally. As for the Finnish troops, it provides limitless highways for their skis, snowshoes, sleds, and snowmobiles. A few years ago my husband and I flew into Kotzebue, on the Alaskan coast, in late April and found Eskimos sad because there soon could be no more visiting of family and friends across the water. The snow was getting soft. Visiting would have to wait three months, until the ice went out in July, and travel by boat became feasible, whereas through the long winter they had zipped over the snow at will by dog team and snowmobile. Visit any interior Eskimo or Athapascan Indian village in spring and you are almost sure to find the men hard at work hauling firewood. "Soon the snow will go and we won't be able to get in here to cut the trees," they will tell you. At the same time, if the top of a knoll or the village airport has lost its snow and wind is kicking up sand from the exposed ground, women may complain that "today sure is no good for drying meat." They may even be hoping openly that their men don't find caribou until the wind stops or a fresh blanket of snow has been laid; else how can they keep the meat clean?

In the past, northern people knew all the varying attributes of snow and employed them. For example, pieces of snow are helpful when you're thirsty, not for eating, but to use as blotters to soak up water from spots that can't be reached directly with lips or a dipper. A good place to look for such water is close against rocks that jut up from a frozen pond; such rocks absorb heat from the sun and melt the ice in immediate touch with them. Or, if you fall through ice into a lake or river, or the ocean, snow can draw the water out of your clothing before it wets through to your skin. All you need do is roll in a snowbank immediately. Cold dry snow will freeze the liquid water onto its crystals and solve your problem.

Snow is remarkably varied. For example, consider hue. There is the glistening white of wet spring snow and the dull powdery white of new-fallen dry snow, the glitter of sequin crystals following a bitterly cold night, and the forlorn grime of snow soiled by city dirt and country dust. Snow can be gray and invisible against the horizon on an overcast day, or it may turn violet-blue as the rays of incoming light are filtered out except for the blue end of the spectrum. Frank Debenham, a geologist in the Antarctic with the explorer Robert Scott, once told of showing the expedition ornithologist, Bill Wilson, a polar landscape he had painted in tones of dull white. Wilson, himself an accomplished artist, asked, "Is that what you really saw— white snow? It's very rare, you know." Thereupon Debenham stepped outside and realized that it was not an all-pervading whiteness that dominated his senses. Instead the snow was shadowed and textured, reflecting tints of the sky.

Usually the exact hue of snow is a matter of lighting. But it can be built in. Pink snow regularly patches the alpine country of western North America from California's Sierra Nevada to Alaska's Brooks Range, and it is scattered throughout the world's high elevations and high latitudes. In this case the color comes from the chilly counterparts of hot-spring algae, well known for their brightening of steaming pools at Yellowstone and Mount Lassen. Algae are among the planet's most adaptable life-forms, surviving from nearly boiling to well below freezing. Of species living their whole cycles in snowbanks, the most abundant forms tint the snow from faintly rosy to decidedly watermelon-red, depending on concentration. At times such snow even takes on the aroma of watermelon (although not the taste). Other forms of algae turn snow yellow or green, and at least in Europe one type turns snow blue.

Aristotle centuries ago spoke of colored snow, but detailed study waited until fairly recent time. The arctic explorer John Ross perhaps laid the foundation in 1818 when he noticed cliffs of red snow edging Baffin Bay and collected samples. He took these back to England, but astonished biologists couldn't agree on what might be causing the odd coloring. A few decades later their successors recognized blooms of microscopic algae as the reason, and today more than one hundred species of algae have been identified in snow and mapped geographically. Some of these are phenomenally catholic in their acceptance of habitats, growing in seawater, soil, and hot springs, as well as in snow. Others are true cryophiles, requiring cold conditions. Several even endure being frozen for years at a time without losing viability. Under normal conditions the pigment of these algae absorbs enough solar energy to melt the immediately surrounding snow. This provides the merest film, of water, but it is enough for metabolism. When winter comes many species of snow algae seem to hold over in a dormant state, usually within the soil under snowfields. When spring brings the right combination of overall melting and light, special cells equipped with hairlike flagella quickly develop and swim up through the snow to its surface. The plant's active, summer cycle then commences.

Investigators now find a whole ecological chain of minute life-forms in snow, dependent ultimately on algae as the bottom of the food pyramid. For instance, the snow worms that Robert Service immortalized are real. Living at Mount Rainier, Washington, where my husband served as park ranger, we occasionally observed the conversion of colleagues from skeptics in this matter into believers. In one case a ranger who was stationed close to the six-thousand-foot level of the mountain, where snow usually lasts into August, customarily chilled drinks by reaching out the window for handfuls of snow. He did this with a ritualistic flair intended to impress guests with the romance of ranger life. Then came an evening of sad truth when worms were found in the bottoms of the glasses!

Worms, as algae, are common in snow and ice, several species of them strictly limited to such a habitat. All belong to the one genus *Mesenchytraeus,* a distant relative of earthworms. Sometimes these worms congregate by the thousands, writhing at the surface of a snowbank like so many fine, inch-long bits of thread. Black is the most common color, although some are brown. The pigment evidently lets the worms melt their way up and down through the snow by absorbing heat from the sun. Algae supply food through the hot months and when cold comes the worms vanish, as well as the algae. Probably they, too, go beneath the snow, although nobody knows positively. Springtails, or "snow fleas," are another form of small animal life that thrives in snowfields. There also are minuscule rotifers and protozoans, a decomposer chain of bacteria and fungi that breaks down the organic debris of all the life-forms, and a predator chain including spiders that feed on the snow fauna and birds that feed on the spiders. Sterility, often associated with snow, really isn't one of its inherent characteristics.

Neither is the common assumption that snowflakes never duplicate one another necessarily true. No dictum of nature keeps all snowflakes from being identical, however renowned their variety; and, since no physical force imposes differences between them, mathematical odds probably favor duplication. Such twins haven't been reported, but the sheer number of flakes that fall and their fleeting existence as separate crystals augur against enough observation to say for sure one way or the other.

Vincent J. Schaefer, dean of United States meteorologists, calculates that it takes more than one million crystals to blanket a two-square-foot area with snow ten inches deep. Multiply this by the nearly one quarter of the world's land surface that is whitened each winter and by a total annual snowfall that in places amounts to two hundred feet or more, and the possibility of duplication—if not the certainty of it—becomes impossible to deny. Having thus computed volume, next add to it the number of snowflakes that reach the earth yearly and also of those that turn into rain on the way down; then multiply this by the eons snow has fallen, and the odds favoring duplication rise still more. Yet the wonder of snowflakes' formation and pattern remains undiminished whether or not a crystal caught this year on a gloved finger may possibly have had an identical counterpart that millennia ago became part of the Greenland ice cap or of an Eskimo igloo.

If not infinite, the variety of snow crystals is nonetheless great—and correctly speaking, snow crystals should be distinguished from snowflakes. A flake is an assemblage of individual crystals, both whole and broken, joined together in falling. Such flakes may be as much as an inch, or even two or three inches, in diameter. They form only at relatively mild temperatures. The polar regions, for instance, never receive snowflakes but only separate crystals, usually so fine and simple as to be virtual snow dust.

The Russian meteorologist Shuchukevich, studying snow near Leningrad, reported in 1910 that he had noticed 246 different kinds of crystals during 176 days of falling snow. A single day produced thirty-seven distinct categories. Ukichiro Nakaya, working in the 1930s on Hokkaido, the north island of Japan, one year made a point of taking samples throughout every snowstorm all winter. Even when the thermometer registered far below zero, Nakaya and a colleague, Y. Sekido, worked nonstop so long as snow was falling. They would catch twenty or thirty crystals on a piece of glass, note their forms, then quickly study and measure each one under a microscope. Five or ten minutes later they would catch more crystals and repeat the analysis. In this way they found that usually one type of crystal dominated only a few minutes before some other kind began first to mix in, then to take over, and soon, to be replaced by still another. Sixty-five microscopic observations during a single ten-hour snowfall revealed fifteen kinds of crystals. In all Nakaya identified seventy-nine categories of snow crystals and still added a miscellaneous category to accommodate mavericks.

In 1951 the International Commission on Snow and Ice proposed a classification scheme recognizing seven basic forms of falling snow crystals plus ice pellets, hail, and graupel (crystals heavily coated with rime). These seven are star, plate, needle, column, column with a cap at each end, spatial dendrite, and irregular. Partly to sort out the "irregulars," C. Magano and C. U. Lee a few years later classified 101 types of snow crystals, making fine distinctions such as hollow bullet, solid bullet, column with dendrites, cup, pyramid, hexagonal plate, stellar crystal with plates at ends, and so on. Of these the stellar crystals are the ones most commonly known. They are assumed the most typical and are the form behind the "no two alike" belief. Beyond question such stars, often perfect and intricately patterned, rate as the most beautiful of all. But they also are not the most common. Irregular, unsymmetrical crystals and aggregates formed into flakes are far more typical.

Wilson A. Bentley, an American farmer-photographer, a century ago gave the public its first magnified look at the exquisite intricacy of individual snow stars. His photographs still are used. As a young man Bentley had combined a camera with a microscope and worked through the cold of Vermont winters to capture the beauty of snow crystals. For forty years he photographed, producing over six thousand photomicrographs, one-third of which were published in 1931 in a book called *Snow Crystals,* which included a brief text by W. J. Humphries. Unlike Shuchukevich or Nakaya, who both approached snow strictly scientifically, Bentley selected his subjects largely on an esthetic basis and even retouched some negatives to enhance the crystals' beauty. His results, thus, are more selective and idealized than truly representative of snow as it falls. What form a crystal will take depends largely on temperature and the availability of water vapor. Bentley knew this and commented that the crystals he concentrated upon fell only under certain conditions. But the beauty of his images overshadowed the truth of his words so far as the public was concerned, and, seeing representations mostly of flawless stars, popular belief mistakenly began to consider them the most frequent type.

From early times men have pondered the singular nature of snow. Among early Chinese scholars the six-sided nature of snow crystals was attributed to water, and both snow and water were given an association with number six. In Germany the seventeenth-century astronomer Johannes Kepler published a small pamphlet speculating at length on the "six corneredness" of snow, but without realizing its true crystalline nature. About the same time the French mathematician and philosopher René Descartes, who was living in Holland at the time, published the first scientifically accurate drawings of flakes and commented on various shapes in addition to six-sided, single-plane crystals. In 1665 England's Robert Hooke added even greater accuracy of detail by examining snow under a microscope, which was a fairly new device at the time. By the nineteenth century, while Japan still was closed to the West, the feudal lord Toshitsura Oinokami Doi published eighty-six sketches of snow crystals he had studied with a "Dutch glass" (microscope), and a few years later he issued a second volume of "snow blossom" sketches. During this same general period two Englishmen made firsthand observations of snow in remote regions. William Scoresby, a whaling captain, published a book of experiences called *Arctic Regions* in 1820; and James Glaisher, founder of the Royal Meteorological Society, made balloon ascents as high as thirty-seven thousand feet and published his findings in 1855, including splendid drawings of snow crystals.

It is known now that six sides are basically inherent in the atomic structure of snow crystals, although some are three-sided or five-sided. Form and growth rate are the product of environmental conditions. At a given saturation of air and with a certain type of nuclei present, hexagonal plates form at temperatures from freezing to 27° F.; needles at 27 to 23°; hollow prismatic columns at 23 to 18°; hexagonal plates again, but of a different type, at 18 to 10°; fernlike stars at 10 to 3°; plates at 3 to −13°; and hollow prismatic columns at −13 to −58° F. At least snow-crystal growth works this way in a laboratory cold chamber. Simply lowering or raising a hair within the chamber to move from one temperature zone to another will change the form of a crystal growing on it. Nobody yet knows precisely how or why this happens. There is some theoretical knowledge of it but no complete understanding.

Artificial production of snow in cold chambers has greatly eased study because working only with natural snow often is troublesome. Snow generally is studied at the earth's surface, yet it forms high in the atmosphere and changes constantly while falling and after alighting. It melts and vanishes if exposed to the warmth of usual photographic lights or if brought into a comfortably heated laboratory. To solve this Nakaya undertook the production of crystals artificially, watching every stage of their growth. In 1936 the University of Hokkaido completed a low-temperature laboratory for him, a thirteen-foot room chilled by refrigeration to as low as −58° F. Working in this chamber Nakaya's problem was how to keep a crystal suspended long enough to finish growing, which often takes an hour or two. He tried to form crystals on various filaments such as cobweb, silk, cotton, wool, and rabbit hair, but he got only fuzzy coatings of frost that made the threads look like frozen caterpillars. Then at last, while repeating experiments many times under varied conditions, Nakaya produced a single artificial snow crystal with six rays and a rude symmetry. Eventually he found that the microscopic knobs of rabbit hair most readily act as nuclei for crystal growth.

Schaefer in 1946 also experimented with a cold chamber, by converting a home freezer for the purpose. He lined its chest with black velvet and inserted an observation window and a spotlight. Next he dropped in a damp cloth, which gave off vapor that froze into crystals. The beam of light showed up the entire process from initial condensation onto a nucleus to dropping to the floor of the freezer as snow. Schaefer earlier had developed the use of a white plastic resin to replicate actual snow crystals, for easier study. His technique involved adding a drop of the resin dissolved in ethylene dichloride to a crystal caught on a chilled piece of glass, then waiting for thirty minutes while the solution dried. Water molecules escape from such a treated specimen, leaving only the resin which perfectly duplicates the original crystalline structure. At least it does if all goes well. The glass slide that collects the snow, the needle used to manipulate it, and the resin solution all must be chilled. The drop of solution should be the same diameter as the snow crystal, the crystal needs to be deposited onto the resin flat, and human breath at all stages must be kept from slide, drop, and crystal lest its warmth cause melting or its vapor leave a coating of rime. Nonetheless by this means Schaefer could leisurely and repeatedly study the plastic versions of the snow he produced in his cold chamber.

What happens as snow forms in the clouds probably never can be as fully understood as what happens in a cold chamber, but since the same kinds of crystals are produced under both circumstances certain statements can be made confidently. Snow is unique as a chemical compound capable of a great variety of crystal habits. Its basic structure consists of molecules with one oxygen atom at the center of two hydrogen atoms held by electric charges at 120-degree angles from the oxygen atom. A single snow crystal may have 100 million such molecules, each with this 120-degree arrangement. To start growing, a crystal needs a nucleus with a suitable molecular structure. This may be dust from a farmer's field or a volcanic eruption, exhaust gas from city traffic, salt spray from the ocean, a tiny splinter of ice broken from a falling snow crystal, or even a microorganism. Given some such suitable nuclei, crystals will develop, providing that droplets are present within a cloud of the right temperature or that there is an excess of water vapor present. If these conditions are met, droplets will sublimate onto the nuclei, in the process setting up minuscule currents that catch additional vapor molecules which also whirl onto the surface of the crystal and become part of it. Depending on temperature at the time of nucleation, growth may be dendritic, platelike, or columnar; and whatever form a crystal takes at first its final growth will be determined by temperature, humidity, and barometric pressure, all of which vary with altitude.

In general, large intricate crystals form at relatively warm cloud temperatures when ample moisture is present, and small elementary crystals form at low temperatures when the air holds less moisture. If a falling crystal drops into air warmer than itself, sublimation quickens. Rays shoot out until the crystal warms to the temperature of the new air layer. Then, differences lessened, a new growth pattern starts. Since crystals fall with their broad surfaces roughly horizontal and they usually rotate, their edges are in contact with the most water vapor, and growth takes place primarily there. If a crystal, or a flake, passes through a layer of supercooled droplets while falling, it will be coated with rime. Droplets remain liquid at below freezing temperatures only so long as they stay suspended; they freeze the instant they touch anything solid, including falling snow. If enough droplets coat a crystal it becomes graupel, a tiny soft lump so heavily rimed that the underlying crystal form can't be seen.

Snow changes as it forms and falls, and it continues to change after landing. On the ground snow passes through stages of metamorphism that lead ultimately to glacial ice if conditions are right, but the process may—and probably will—be interrupted at any stage. The fluffy blanket of new snow that falls without wind and softens the winter earth is an unstable emulsion of air and flakes. It is a froth. The larger the flakes, the looser they lie and the more quickly they change. The more intricately branched they are, the larger their surface in proportion to volume, which in itself carries a tendency for the redistribution of mass and energy. A rounding of form results, lessening the ratio between surface and mass. Apparently this occurs primarily as molecules sublimate away from the points of stars, where vapor pressure is greatest, and are redeposited into the notches between rays. The hollows of a crystal thus fill in at the expense of points and edges. Intricate pattern and surface detail are lost, and small round grains eventually result. Each may retain the mass of the earlier crystal, but the form is changed forever. Or, in a similar way, small crystals may lose their molecules to large ones; fine projections have higher pressures than large ones, and so small crystals with the sharpest angles lose molecules to adjoining large crystals with blunter angles. In this way snow settles and becomes denser. At this stage it is called "old snow," a term that applies to the stage of crystal growth rather than to the passing of time.

If metamorphism continues, *firn* develops from old snow. This is a further compaction and hardening, with bonds from grain to grain tighter and mechanical strength and density increased, but with spaces still remaining between the grains. If the process continues still farther, firn becomes glacial ice. Individual grains touch, and air is present only as bubbles within the crystals, not in spaces between them. At this stage what once was snow takes on a vitreous character.

Snow layers close to the ground usually are warmer than upper layers because of heat given off by the earth. The effect of this is that vapor moves within snow from lower and warmer layers to upper and colder layers, and new crystals form. Their shapes may be hollow cups, prisms, scrolls, columns, or hexagons, depending on conditions of growth. Near the bottom of a snowpack, depth hoar may form, a layer so porous as to amount to little more than a latticework of ice walls and space. Depth hoar can't bear much of a load without collapsing and may be highly susceptible to shearing. It is the bane of cross-country skiers, since its hidden layers continually give way; but for the same reason it is the blessing of animals such as northern lynx which easily break into it and feed on voles they hear scampering about at the ground surface. To eliminate avalanches set off by the wholesale collapse of depth-hoar layers, researchers are trying to find a chemical means of control. It might be added to snow to weaken depth hoar and thereby release avalanches while they still would be small, or it might affect the growth of the crystals themselves and produce a more stable layer in zones known to be critical for avalanching.

Of external forces affecting snow, none outranks wind. It may pack falling flakes into a firm crystalline blanket rather than a soft airy one, or it may drift already fallen snow into gentle ripples or sharply pinnacled *sastrugi*. It may build cornices that curl over mountain ridges, or etch graceful saucers around tree trunks and boulders and chalet corners as eddies first spiral particles upward, then bring them back into the main wind stream. Wind mechanically rounds snow crystals by shattering delicate rays and spikes, simplifying their outline and fitting them together more tightly. It also steps up the rate of sublimation, eating away molecules along the points and margins of crystals by evaporation. Wind brings atmospheric vapor to condense onto snow and freeze, thereby enlarging grains and bonding them. It sorts grains according to size and deposits them into layers that vary with density, permeability, and thermal character. Eskimos call snow not yet picked up by wind *api*; what has been reworked by wind and deposited into a firm mass *upsik*; and what moves along the ground surface *siqoq*. Windblown snow is more than simply windblown snow; it is all these variations.

Layering persists throughout the life of a snowpack and even after its transformation into glacial ice. Cores taken at Byrd Station from as deep as four hundred feet within antarctic ice show distinct layers representing centuries. On the Blue Glacier of Mount Olympus, Washington, I remember one night descending a seventy-foot vertical shaft dug through five years' worth of snow layers which, toward the bottom of the shaft, had been compressed into ice. This was at the University of Washington's glacier research station headed by Ed LaChapelle. The shaft was large enough for one person at a time to be winched up or down in an oversize canvas bucket while a fluorescent tube was raised or lowered in a second, smaller shaft separated by eight or ten inches from the main shaft. The tube's glow backlit the ice and disclosed the layering in a manner all the more dramatic because of the night's inherent blackness and one's own awareness of descending through the snowstorms of the past as well as through the inner structure of a glacier.

Slippage along the boundaries between snow layers holds potentially immense force. At Mount Rainier each fall, rangers marked the Paradise fire hydrants with lengths of two-inch pipe set upright to rise above winter's snow accumulation. This permits digging the hydrants free in early summer, while snow often still lies fifteen to twenty feet deep. Well into July, entry to Paradise Inn is through a snow tunnel and fire hoses can be connected to hydrants only by digging into the snow. Following a pipe down, however, never means sinking a straight-sided shaft because invariably snow layers creep downslope at different rates during the winter and their pressures against the pipes twist and distort them. Consequently holes dug to reach the hydrants also take on odd angles.

Living at Rainier impressed us with how much snow varies according to exact conditions. Through the winter, wet snow piled like pillows onto every horizontal surface, from the crossbars and seats of our sons' swings to the rough ridges of tree bark. On cold days it squeaked underfoot as individual crystals rubbed against each other and broke, and on days close to freezing, or above, it silently compressed beneath boot soles and skis. When summer came, climbing the mountain was simplified by crusted snow that would bear our weight, or made harder by crust so icy that the points of crampons and ice axes wouldn't bite in or so soft that it broke beneath each step. Even worse was no crust, when we sank ankle-deep in slush or found our crampons balling up with snow stuck to their spikes. Adhering snow then must be repeatedly knocked off by hitting the side of your foot with your ice ax every other step or so. When one is weary, this added effort can seem too much, and I remember on my first Rainier summit climb that I fell again and again on the descent as my crampons balled up. Sliding free while fallen gave such effortless progress that each time I delayed rolling onto my belly and digging in my ice ax to self-arrest the fall.

From a scientific standpoint a litany of terms describes the differences and changes within snow: allotriomorphic-granular, hypediomorphic-granular; paratectonic perecrys-tallization, regelation, congelation, infiltration, densification. The words precisely describe the variations and continuing stages of state within snow, only slightly suspected by most of us who stand at the top of a ski lift or shovel a walkway clear after a blizzard. Certain properties of snow, ill defined within our minds though they may be, nonetheless evoke characteristic human responses. The urge to slide, for instance, seems nearly universal whether on skis or sleds, or with the simple wooden clogs traditionally soled with metal for the delight of children in northern Japan or the "sliding shoes" that coastal Eskimo parents made for their children (ovals of skin worn under the boots and held with straps).

The ready compressibility and cohesiveness of snow also elicit a predictable response: snowballs. Even those who never before have seen snow may become instant boys. Three African dignitaries once came to Mount Rainier following a summer international conference, their faces elaborately tattooed, their teeth filed and set with diamonds, their robes richly brocaded, their bearing totally regal. Yet the moment they saw snow the air filled with snowballs. Gold slippers wet and robes of state flapping, the battle was on. So irresistible is this urge that Eskimo children in a village along Alaska's Kobuk River whom I asked about their feelings toward snow invariably answered in terms of play. They mentioned sliding, snow fights, snowmen, tunneling in snow, making snow dolls, mounding snow into targets to shoot arrows into, and drawing in the snow with sticks. "We always draw houses and clocks and stoves," one child wrote, indicative of an influence from the outside world which is essentially new within his own generation. Another noted that "when it is cold we can't make snowballs," an awareness basic to childhood throughout northern latitudes and high elevations. Cold snow won't stick together.

Perhaps not surprisingly, animals as well as humans seem to delight in snow. Hiking in the Olympic Mountains one July, I watched a black bear slide down a snowpatch on its belly, then lumber back up and repeat the pleasure. In similar fashion polar bear sows stand to catch their cubs at the bottom of a snow slide and wait indulgently while the young run up to slide again. Otters are renowned for their snow slides, and evidently snowshoe hares also play in snow. When the whitened world gleams in the moonlight they sometimes gather in considerable numbers to bound and leap in a pixilated dance that is apt to end abruptly as the silhouette of an owl glides silently overhead. Even monkeys, not common in snow country, have been observed making snowballs. A troop of 230 macaques brought in 1965 from Hiroshima Prefecture, Japan, to the primate research center at Beaverton, Oregon, so far have this distinction exclusively, although there may well be others that haven't been noticed or reported. Making snowballs began six years after arrival, on January 13, 1971. A male of low status within the troop that winter discovered while eating snow that he could make it into a ball by rolling what he held in his hand along the ground. Intrigued, he pushed and rolled until he had a snowball well over a foot in diameter. Ever since, other monkeys have joined in and now even those who don't make snowballs like to pat them and sit on them. So far no throwing has taken place, only the rolling of snowballs and playing and resting on them. Behavioral psychologists call the whole procedure an outgrowth of "manipulative skill, intelligence, and leisure time."

All who have experienced snow more than once or twice know that some kinds will do for snowballs—or any other specific purpose—whereas others won't. Snow isn't simply snow; it is different kinds of snow. English expresses this poorly. A language developed in a moderate climate, it lacks the subtleties needed to speak adequately of snow. Northern peoples, however, have the words for fluent, concise description. For instance, an English speaker might talk about "a place that has been cleared of snow by the wind," but a Soviet tribal hunter of the Altai or Transbaikal could use the single precise word *vyduv*. English speaks of "an accumulation of snow blown into a depression and likely to remain there well into the summer," whereas *zaboy* says the same thing. Furthermore, these single snow words are more than labels; they carry amplifying connotations. A vyduv is where reindeer graze in winter; a zaboy is where they lie in summer to escape from flies. Siberian peoples have one word for a deep, soft snowfall, a different one for walking in such snow, and still another for the kind of fresh-fallen snow that facilitates tracking animals. Wind-packed snow has its own term; so do the odd pinnacles sculpted by wind. Along the north coast of European U.S.S.R., when large fluffy flakes are falling, people say in one word that "the pad is coming down." But if wind is whipping the snow they use a separate word and it, in turn, is different from the word for snow already on the ground and being blown by wind.

Probably the greatest linguistic finesse of all belongs to the Eskimos whose lives are so filled with snow that they phrase the question "How old are you?" by asking "How many snows-there-is-none have you seen?" Their language— so similar around the entire arctic basin that a seal hunter from Greenland can understand one from Alaska—distinguishes two dozen or more kinds of snow. The basic kinds of windblown snow are only a beginning. There also is new-fallen, soft snow; fluffy deep snow that fell without wind; wind-packed snow firm enough to walk on; wet snow belonging to springtime; the walk-anywhere crust of an early morning that comes after a cold night; and the same crust after it has thawed to corn snow but is still fine for traveling; dry and sugary snow; snow for igloos; long tapered drifts in the lee of objects; rounded smooth drifts; snow lying on the slope of a hill; local ground drift; extensive ground drift—and more, more, more.

Ironically, snow brings warmth of a sort, for the earth's white "blanket" is literally that, a blanket. It holds in the warmth of the earth and wards off the frost of winter air. This affects the wild plants of the tundra and taiga and alpine meadows, and it presents an immeasurable boon to northern farmers. Where snow is lacking across the prairies and steppes, soil regularly freezes to a depth of several inches. One Russian comparative study showed that with a six-inch blanket of snow, frost penetrated less than an inch into the soil, whereas adjoining snow-free fields were frozen for more than a foot. Furthermore, snow not only insulates against low minimum temperatures but also against fluctuations. A bare soil surface may be heated by direct sunshine, then cooled in seconds as clouds pass overhead, and its temperatures often surge drastically from day to night. Snow moderates these swings and that can be a crucial advantage, depending on plants' hardiness, for winterkill and stunted growth can occur without actual freezing.

Surprisingly, snow also benefits plants directly by supplying them with fertilizer. One investigator places the value of this to prairie farmers at around twenty dollars an acre in nitrates spread on the fields by an average winter's snow. French peasants have a proverb to the effect that a proper February snowfall is worth a pile of manure. Sulphate, calcium, and potassium, as well as nitrates, are delivered to the earth by snow (and rain). These ions—and others in lesser amounts—come from ocean air masses, dust raised from the earth, atmospheric gases, and industrial pollution. Their presence has been studied as far back as 1852 and regarded as of possible significance as a source of nutrients for plants and for river and lake organisms.

Not all growth fostered by snow is a blessing to farmers, however. Snow mold is a curse throughout the cereal belt of the world. If snow falls early, especially on unfrozen soil, certain fungi may become active enough to menace young plants. Even if the plants are only moderately affected, they may lose so much nitrogen from blighted leaves that they respond slowly to spring warmth, which ultimately increases problems of weed control and delays harvesting. In a recent winter, wheat losses from snow mold averaged forty-five hundred dollars per farm in one county of eastern Washington. Snow favors the growth of fungi by holding soil surface temperatures close to freezing and maintaining a high humidity. Ample oxygen is present even when snow lies deep, and since fungal growth thrives on dim light anyway the near-darkness doesn't matter. A foot and a half of snow has been found to reduce light by 99.5 percent, which is ideal for snow mold. Green plants can't photosynthesize under such conditions, however, and the longer they go without manufacturing food the more their leaf proteins break down, weakening them and making them susceptible to disease.

Hastening snowmelt by spreading whitened fields with fine peat, wood ashes, lamp black, powdered graphite, or coal dust mixed with talc is one way to reduce the danger of snow mold. But to be effective, whatever material is used should be insoluble in water and coarse enough to stay on the snow surface and ride it down as melt takes place. If it dissolves or sinks into the snow and disappears, it is useless. Dusting from planes makes this method of speeding up snowmelt feasible, but the process is a gamble at best. New snow may bury a freshly blackened surface, or a warm spell may melt a shallow snow cover and render the dusting just that much money wasted. Furthermore, any speeding of melt to control fungus must be balanced against its effect on soil moisture, for snow acts as a natural reservoir. In fact, pamphlets for farmers on how to accumulate snow on their acreages far outnumber those for melting it ahead of schedule.

Drift fences and barriers of vegetation with strong, flexible, close-set stalks can control where snowdrifts form and how deep they get. Without such management, snow may blow and be lost, carrying away a major source of water. A four-year test in Montana showed average accumulations of six inches of snow (equivalent to almost two inches of water) where no barriers were employed, and twenty inches where there were barriers. They cut wind one foot above the surface by more than 80 percent at a distance of five feet. At twenty-three feet, wind still was cut by one-third. Such reductions cause drifting snow to settle onto the ground. Even low stubble catches about four times as much snow as fields left fallow.

Snow can be manipulated: Where it is to fall and when it should melt can be decided by man to a considerable degree. To some extent even glaciers—the fossil snow of past millennia—can be pressed into our service. The potential of this control is enormous. Three quarters of the earth's fresh water is stored in glaciers, the equivalent of about seventy-five years' worth of snow and rain for the whole globe at current rates. By far the greatest glaciers are located in the polar and subpolar regions—remote, yet perhaps not beyond reach of technological man if agriculture or industry becomes thirsty enough. At present southern California researchers, financed by the National Science Foundation, have proposed towing icebergs north from Antarctica to meet water needs in the Los Angeles basin, and the Saudi Arabian government has asked French engineers to work out the feasibility of towing antarctic bergs through the Indian Ocean and the Red Sea to the port of Jidda.

The basic scheme already has been put into action. As

early as 1900, bergs were hitched to steamships and taken
to arid Peru; but the undertaking was relatively small-scale
and short-lived. The California and Saudi proposals are
more elaborate. One calls for lassoing icebergs with cables
and linking them into trains as much as fifty miles long to
be towed by atomic-powered tugs. Another recommends
towing single, huge bergs one at a time. Antarctic bergs are
fairly smooth and manageable in shape, and as big as 150
miles across and 1,000 feet thick, an incredible volume
even allowing for melt on the long trip north. (And melt
would be minimized by covering the ice with plastic quilts
almost two feet thick, calculated to hold the loss to no more
than 10 percent of a berg's volume.) Tugboats of the type
used to haul oil-drilling platforms would tow the icebergs,
probably making only about one knot an hour. This would
mean an eight- or ten-month trip from McMurdo Sound to
Los Angeles, about six months to Jidda. On arrival close to
port, the bergs could be tethered and quarried by a scoop.
It would feed great chunks onto a conveyor belt for delivery
into a submerged pipe connected to the shore. Stockpiled
in cold storage on land, the ice then could be melted as
needed. Its water would be the purest available to any
major city, and the cost for Los Angeles should be about
one-third of what the city now pays for water brought by
aqueduct from northern California. For the Saudis the
cost would be far less than that of desalinization.

Even without towing icebergs to the temperate zones,
glacial ice can be managed for its water yield. This is done
regularly in China, Japan, the Soviet Union, and Chile. The
process is the same as that for minimizing mold by melting
the snow from wheat fields, namely the spreading of a
darkening substance. Tests in the Soviet Union have demonstrated that sand spread onto snow can be especially effective when mixed with Prussian blue or black aniline dye. In
the Nagano Prefecture of Japan, power company officials
have used carbon black to melt snow and increase the water
supplied their turbines. Reports credit this blackening with
a stepped-up production of nearly 400,000 kilowatts per
year at one power plant.

The reverse of speeded-up melt also is possible. Melt can be slowed by spreading a thick layer of any insulating material onto the ice, or a coating of some reflective substance can be used to mirror the sun's energy instead of absorbing it. Another technique, being studied in Russia, is to send smoke billowing out over a glacier to block incoming solar rays. Also in Russia and in China, propitious clouds are seeded specifically to increase snowfall and restore glacier wastage, and wind baffles are installed to direct the accumulation of snow into desired areas.

In a long-range sense ancient snow, compressed into glacial ice, constitutes the major potential water source for a world faced with a probable doubling of population within the next two generations and a need of water for everything from drinking and bathing to growing corn (ten thousand gallons per bushel) and producing steel (sixty-five thousand gallons per ton). On a current basis, without stepping up ice melt, our yearly blanket of fresh snow constitutes a sort of placental support for humanity linked by aqueducts and irrigation ditches and transmission wires from the mountains to cities and fields. The wheat that grows in eastern Washington and the electricity used by Puget Sound communities depend on snowfields in Canada, which give birth to the mighty Columbia River. The lettuce from southern California and the cotton from adjoining Sonora, Mexico, and even the bright lights of Hollywood, are the products of snow that falls in the Rocky Mountains and flows through the desert as the muddy water of the Colorado River. The petrochemical industry sprawled from Tokyo to Osaka, the steel mills along the Rhine River, the cattle ranching and the flocks of wild flamingos at the mouth of the Rhone— none could exist if it weren't for the snowfields of distant slopes.

Recognition of these ties led to an International Hydrological Decade, which closed in 1974. Its purpose was global assessment of water sources, needs, and prospects, and its proceedings read like a world gazetteer. A British expedition agreed to look at Patagonian ice and present-day snow. The Japanese began an intensive study of the Nepalese

Himalayas. An American team initiated an inventory in Afghanistan. Iceland introduced an "adopt a glacier" program whereby climbers venturing into the back country were urged to map glaciers and lakes along their routes. The Soviet Union promised to gather enough data to fill "more than 100 volumes." Uganda at the last minute canceled its participation, which would have been an assessment of the Ruwenzori Mountains. The People's Republic of China declined involvement.

Until recent years nobody dreamed of harnessing glacier water, but the concept of surveying winter snowfields for their probable moisture yield originated decades ago. Seemingly this first was done in Europe in the late 1800s, although the record is sketchy. In any case, Charles Mixer, an American engineer working on dam and power projects in the state of Maine, developed the idea on his own in 1900 and thereby fathered a beginning of today's sophisticated snow surveys. Mixer wanted to know the prospects of flooding from spring runoff, and of lakes filling, so he decided to sample snow for its water content. He used a cylinder pushed to the ground and sealed at the bottom by first shoveling out around the tube and then slipping a sheet of metal under the opening. By this means a column of snow could be lifted out and melted to measure its water equivalency.

A second, separate, much better-known start of snow sampling came a few years later, and it most directly laid the foundation for today's system. Oddly, one man's personal exuberance for the winter mountains got events under-way—and he wasn't an engineer cannily assessing possibilities but a professor of Greek and Latin: Dr. James Church of the University of Nevada. Mount Rose, a few miles south of the university campus in Reno, drew Dr. Church like a magnet. In 1895 he and a friend celebrated New Year's Eve by climbing the 10,800-foot peak equipped with "nothing but rubbers and one pair of webs [snowshoes] for the two of us." Mountaineers today might well call such an ill-prepared ascent irresponsible, but at the time no communications or rescue networks existed to concern themselves with other people's routes and itineraries. At Christmas six years later Church and his wife spent a week in a cabin at the 9,000-foot level of the mountain, each evening setting a dead pine afire to let friends below in Reno know that all was well. In 1906 a winter climb of Mount Whitney in California drew him. Whitney, at 14,495 feet, is the highest peak in the United States outside Alaska.

Personal enjoyment—adventure—prompted these treks into the snowy heights, but after Church returned from Whitney he talked with power company officials in Reno about putting his obsession to work. They wanted information on the upper reaches of the watershed; he offered to climb Mount Rose each month and take temperature readings. He had no particular instruction in such an undertaking, only some experience in climbing through snow and a boundless enthusiasm. "We had no training other than that required of an Arctic explorer," Dr. Church was to write later, "*viz*, the willingness to endure and the desire to observe." That was enough. The power company thought the proposal worth a try, and Church arranged for horses to haul a small instrument shelter over the snow to the summit of Mount Rose. He and a colleague then commenced biweekly ascents, leaving sleeping bags hanging from the limb of a giant pine at nine thousand feet, so that they could sleep there one night, then climb on to read the instruments the next morning.

Once they tried three times within a week to get the readings. The first night they dug a trench into the snow to protect them from winds of seventy miles an hour and zero temperatures, and in the morning they struggled on. The instrument shelter was frozen shut so tight, however, that they couldn't open the door to get the record sheet. Back to the pine tree and the trench they trudged. Determined, Dr. Church started up alone a day later, climbing in such heavy fog that he could find his way only by shouting and listening to the reverberations of his voice from the cliffs, or by walking along ridges where the wind was sweeping up and "carrying the soft snow into the air in density and blindingness like flour in an overturned flour mill." Then for a moment visibility returned, and Church saw he was far off course, whereupon he untied the little sled he packed on his back and, lying on it, dropped quickly back downslope and returned to camp. Two days later the clouds lifted and the original party made the ascent, "even leaving their coats behind."

The scientific community and the press followed the readings from the mountaintop with interest, for at the time few specifics of high-elevation conditions were known. Before long, money came in for improving the Mount Rose Observatory and the pioneering snow survey efforts entered a new phase. The sleeping-bag tree was upgraded into a little hut half-a-man high and three-men wide and furnished with a "cooking lamp and an immense rabbitskin blanket." From that point of origin, the methods developed by Dr. Church became the standard not only for the western United States but for the continent and abroad in nations such as Norway, Sweden, Switzerland, India, Chile, and Argentina. His Mount Rose Sampler, as first designed, consisted of steel tubing in lengths of up to ten feet. These could be screwed together to reach through a snowbank and extract a sample of snow from its entire depth, clear to the ground. Through the years various aspects have been modified. Aluminum tubing, fitted with steel cutting teeth, now reduces the weight by about half, and tube lengths are reduced to thirty inches for ease in handling. Fiber-glass and plastic linings minimize sticking and simplify getting samples from deep, dense snow. Oversnow vehicles have been added to skis and snowshoes as a means of running snow courses, and telemetry, helicopters, and satellite photography now increase still more the reliability and sensitivity of snow readings.

Surveys have become a matter of men closeted in warm offices surrounded by electronic scales and calculators that are recording the conditions of distant, cold slopes. One system relies on a series of stakes set out in snow courses for which long-range data already are known. These are photographed from the air to provide readings of snow depth for comparison against the overall mean. Another system involves metal pillows that register the weight of snow falling onto them. A third relies on gamma emissions from lead-shielded packets of cobalt-60 set at ground level before the first snowstorm of the winter. A counter suspended above each packet measures the rays emitted by the cobalt and beamed straight upward, and, because water impedes their transmission, the resulting electronic pulses can be correlated to the moisture content of the snow. Canadian, American, Soviet, and Norwegian tests have established that naturally occurring gamma emissions of various other radioactive elements can be read by airborne spectrometers and used for determining the water equivalency of snow. Or, airborne microwave radiometers can be used to sense the profile of ground snow cover, an effective means since liquid water content directly and strongly affects microwave emission from snow.

Satellites orbiting the earth now map snow cover and, although depth and moisture content can't yet be detected by this means, what they tell about the extent of snow has proven extremely accurate both in mountainous terrain and across flat land. This is true despite the interference with some readings from clouds and thick canopies of coniferous forest. Radar can be used to estimate the water equivalency of snow while it is falling and to map the thickness of glacial ice. Seismic waves also measure glaciers from their surface to bedrock, and ice buried beneath the rock and gravel of a moraine can be detected by how it radiates infrared energy from deep within the earth.

Myriad complications interfere with all of this modern wizardry but, even so, its cumulative accuracy is relied upon for an ever-escalating range of applications and, rather than tasting triumph in their ability to read the hitherto undecipherable, today's forecasters more nearly feel concern over the possible consequences of misinterpretations. Should ski resorts prepare for a severe avalanche season? The answer lies in assessing the amount of the snowpack and its nature. Will there be water enough for farmers, or should they reduce their planting? What of floods? And what of the prospects for hydroelectric power generation? How about increased roof load on buildings throughout the snow belt? Is there a threat of serious pollution if spring melt lags behind normal and fails to dilute adequately the effluent carried by rivers? Such pollution has occurred. The social and economic impacts of snow surveys are crucial for the technological world.

Increasingly also, men today are served by information from the snows of the past. Lying layer on layer within glacial ice, they can be "read" for a surprisingly large amount of baseline data with which to compare our present conditions with those that have gone before. Volcanic eruptions through the centuries have left signatures in the form of ash fallen onto what at the time were the surfaces of the snow. Ice cores taken from Antarctica revealed two thousand such ash deposits during the last seventy-five thousand years. (The core samples spanning this much time totaled more than a mile long.) Radioactive snow gives incontrovertible proof of nuclear explosions in the atmosphere. Extraterrestrial fallout also shows up in glaciers—part of the more than nine million metric tons of micrometeorites that shower the earth each year at present rates. Most of these tiny fragments are hopelessly difficult to differentiate from the earthly dust they fall onto, but not so with the fragments that land on snowfields. They show plainly, and continue to do so as layers of fresh snow and of additional micrometeorite showers build on top of them.

A possibly alarming increase in the lead content of the environment has been noticed by ecological sleuths analyzing the icefields of Greenland. It correlates with human contamination of our planetary nest. A fivefold increase in lead shows up for snow layers dated from 800 B.C. to A.D. 1750, and another fourfold increase is recorded from 1750 to 1940. The percentage rise since 1940 is even sharper. The figures reflect man's activity beginning with the Industrial Revolution, particularly widespread coal burning and metal smelting, and also a wholesale use of leaded gasoline in automobiles since World War II (most of it attributable to the United States alone). Analysis of snow layers within antarctic ice presents a different picture. Even recent lead levels there so far are scarcely above those of ancient ice, and the difference from one hemisphere to the other seems to be the concentration of industrial activity in the north.

Fossil weather and climate also are recorded by the world's glaciers. Snow that was blown about by wind when newly fallen shows today in layers of rounded, compacted grains. Ancient crusts that formed as the sun melted the snow surface can be identified by condensation along the bottom of a layer and an evaporation zone directly above it. Summer can be distinguished from winter in regions where dust blows onto snow, is concentrated by melt, and then covered with fresh snow when winter returns. Isotopes within snow point to climate characteristics through time, a chemical indicator based on variation in the numbers of protons and neutrons within the nuclei of certain elements. Deutrium, tritium, and oxygen-18 are particularly useful tracers. Air held by bubbles within ice can be vacuum extracted and its carbon dioxide dated by radiocarbon methods to determine how long ago that particular snow fell. Pollen grains can be recovered and identified to yield clear information regarding past climates, since different types of plants grow only under certain conditions.

A large portion of the earth falls within the realm of snow. If the cold regions are defined as where the coldest month averages freezing or below, then about half of the land in the northern hemisphere is included. For the United States, the line angles down the Cascade Range of Washington and Oregon and continues slightly west of the California-Nevada border, then about at Reno starts wavering across the continent between the thirty-fifth and fortieth parallels. The spectacular red sandstone country of northern Arizona and New Mexico is included. The Texas panhandle and Arkansas are barely inside the line, which then continues eastward across southern Kansas, the middle of Missouri, southern Illinois, Indiana, and Ohio. From there it follows below the Pennsylvania border and goes out to sea beyond New Jersey, New York, and Connecticut. All of New England is within the cold region, both coastal New England and inland.

In Europe the line strikes southward along the fjord coast of Norway, runs along the German and French border, follows the Alps practically to Monaco and eastward across the top of Italy, then swings down through Yugoslavia, across Bulgaria to the Black Sea, on across Iran, north India, Tibet, and into China just north of Shanghai. Japan lies largely outside the line, except for the northernmost island of Hokkaido plus Honshu from Sendai, north. From the Asian coast the line crosses the Pacific to the Aleutians and starts south along the Alaskan panhandle and the British Columbia coast. Plotted along a zero-degree Fahrenheit line, instead of a freezing line, this northern cold region leaves out all of the United States except for a narrow arc of North Dakota and Minnesota. Europe is entirely out, and the line enters Siberia from the Arctic Ocean following the Ural Mountains, at sixty degrees north longitude. This puts the whole of Siberia inside the line (except for the Kamchatka Peninsula), and Mongolia and Manchuria also are included.

Almost equally vast regions are delineated by plotting the realm of permafrost, permanently frozen ground which causes soil to swell and settle, and buildings to cant off plumb and highways to buckle. Permafrost reaches as deep as one thousand feet into the earth in places, and where it does facilities as basic as a water supply become major problems. A deep well may be mechanically unsound, and a shallow one will freeze up. Pipelines cannot be exposed to freezing, and they also must not be allowed to transfer heat into the ground and disturb its thermal regime. The upper ground, which may freeze and thaw each year, will change from rock-hard in winter to oozing mud in spring, and during fall freeze-up stresses as great as fourteen tons per square inch can be expected.

Still another way to define the cold regions of the earth is by snow. It blankets half the land surface of our planet at least temporarily and 10 percent of the ocean surface. About forty-eight million square miles of the earth's surface lie under a constant blanket of white; about six million square miles are ice-covered, the rest snow-covered, sometimes deeply so.

Within the United States the record for deep snow belongs to Paradise, situated on a mile-high shoulder of Mount Rainier, where the winter of 1970–71 brought 1,027 inches of snow. Even in nonrecord years winter entry into the patrol cabins throughout the high country of the park is by digging down to a second-story window. The next heaviest snowfalls recorded in the United States have been at Thompson Pass, Alaska, with 974 inches for the 1952–53 season, and near Lake Tahoe at Tamarack, California, with 884 inches in 1906–07. In the single month of January, 1911, Tamarack registered 390 inches of new snow. In a single storm, from February 13 to 19, 1959, the Mount Shasta Ski Bowl in northern California was covered with a 189-inch snowfall, and in one 24-hour period, during April 1921, Silver Lake, Colorado, received a 76-inch blanket.

Such record amounts, however, have relevance only for our time. In broad perspective they are as fleeting as the snowdrifts themselves, for climate doesn't hold steady, and the land that lies white and cold today may well have grown tropical plants in the past and will do so again. Major snow celebrations as diverse as Alaska's 1,150-mile Iditarod sled-dog race from Anchorage to Nome, and Japan's festival of snow sculpture held in Sapporo each winter, belong strictly to the present, from a climate standpoint as well as culturally. So do everyday events, from prairie farmers tying a rope between house and barn to prevent getting lost during blizzards to snow-country mothers the world around endlessly dressing their tots for snow play, then a few minutes later undressing them. For Buffalo hookers and Seattle skiers— and Bronx Zoo reindeer—the pattern of expected snowfall is based on a period of time so brief as to be indistinguishable within the full flow of earth history. And it is this broad time scale that determines present-day climate.

It is one we only have begun to grasp.

CHAPTER

2

THE CLIMATE PENDULUM

FREAK STORMS FROM TIME TO TIME DUMP SNOW ONTO DIS-
mayed Floridians and whiten the lowland deserts of Arizona and California. A February 1887 snowfall is on record
as having blanketed San Francisco with four or five inches,
resulting in arrests for snowball attacks on cable cars.

Such storms may be one-time occurrences, or they may
portend change. Many signs and many years of data have
to be combined before a trend can be read accurately.
Recently climate has shown signs both of warming and of
cooling. Sharks have been reported off San Francisco, and
cod and haddock, formerly belonging to the Grand Banks
off Newfoundland, have been taken off Greenland. Rose-
breasted grosbeaks, Baltimore orioles, Canada warblers, and
other songbirds have extended their ranges northward; so

have mammals such as coatimundi, ringtail cat, jaguarundi, and opossum. Farmers in southern Ontario a few years ago began to experiment with cotton, Icelanders to plant oats, and New England lumbermen to lose their stands of white birch. Why? Because of worldwide warming—which apparently now may be ending, for other signs point to the onset of cooling. British farmers have seen their growing season decline by about two weeks since 1950, with a resultant serious loss in grain production, and in 1976 more Soviet Union wheat fields were white with snow by October than anybody could remember having happened before.

Greenland furnishes a classic example of overall swings between opposite trends. Indication of recent—seemingly ending—global warming was detected there beginning about 1900. The general whiteness of the landscape began to show a widening coastal rim of green, and sheep and cattle again could be pastured, as they were centuries ago. Records from that earlier time suggest that the name "Greenland" was meant literally, even if with some overstatement. Eric the Red, the famed Viking outlaw, is the one responsible. Exiled from his Norwegian homeland because of murder, he sailed to Iceland in the late tenth century. He joined its Norse colony and again committed murder and was ousted. There was no turning back, so he sailed westward. It may have been rumors of land in that direction that drew him, or maybe he saw Greenland's mountains. In clear weather they appear to westbound mariners before Iceland's peaks have vanished astern. Either way, his voyage was daring, although the northern climate was relatively warm at the time, and the sea probably had no ice.

Eric's thoughts were on settlement, and he is said to have believed that a "fair name" would win more recruits than one lacking built-in promise. He chose Greenland, and his idea, or the skill with which he went about promoting it, paid off. In the year 982, twenty-five ships set out for the new land carrying a probable six hundred or more people plus horses, cattle, sheep, and household and farmstead goods. All went well, and the fleet landed according to plan. Within a few decades Christianity arrived, and soon men-

tion of Greenlanders appeared in the records of Rome as
well as in the ancient Icelandic and Norwegian sagas. By
the twelfth century the population of the colony had swollen
to three thousand. Tithes were being paid faithfully to the
church, and reports indicate a regular export of butter, wool,
hides, thongs, and ivory from Greenland to Norway. Also,
regular yearly expeditions were sailing to North America to
cut timber. (Indeed, archaeologists recently have excavated
a Nova Scotian settlement identified as Viking.)

For about three hundred years the northern settlements
prospered. Then in 1294, the Norwegian king sold a Bergen
firm of merchants the exclusive right to trade with Greenland, even forbidding the colonists themselves to build or
sell ships of their own. Immediately trade began to decline
and in short order years were passing between the arrival
of vessels from Iceland or Europe. Greenland's economics—
and surely also its morale—sagged woefully. As if to compound the trouble, increasing coldness both lessened the
success of farmers and dimmed the enthusiasm of the Bergen
traders. The merchants reasoned that the voyage had grown
too hazardous with the climate worsening. Ivory from Africa
and furs from Russia had become available, and they offered
a greater trade enticement than Greenland's. Furthermore,
both the dreadful Black Death epidemics and political in-
stability were ravaging Europe. Contact with the north
dropped off, and the Greenland colonies were left substantially on their own. A Vatican record of 1347 mentions re-
ceipt of ivory from Greenland parishioners to help in financing the Crusades. After that the written record falls virtually
silent. By the time of Columbus' first voyage a bishop of
Greenland still was listed among the clerical hierarchy, but
there is no hint that he ever sailed north to take up his
duties.

Archaeological investigation in Greenland fills a few of
the gaps in the manuscript record. Excavations in the frozen
earth of a cemetery at Ameralik Fjord reveal that the colonists must have remained in touch with Europe into the
late 1400s; the dead lie buried in clothing of a style fashionable among Europeans of the period. They wear smocks

with tube cowls and cylindrical caps such as are portrayed in medieval paintings. Even more telling are the skeletal remains themselves. They suggest that the contact may not have amounted to much more than clothing style by that time. The bones indicate a people poorly built and deformed. Degeneracy evidently had crept in, probably owing to extensive inbreeding and inadequate diet.

What ultimately happened to the Greenland Norse may never be known fully. The last of the colonists may have intermarried with Eskimos or have been killed by them. They may have died of their own genetic weaknesses. Perhaps they rebuilt boats and journeyed to some other coast or traveled there by sled on the winter sea ice. These possibilities, however, would require a stamina that seems to have been lost. The burials become increasingly shallow from earliest to most recent. Permafrost must have inched closer to the surface as the climate cooled and digging the graves probably became noticeably more difficult every few years. This coming of the cold spelled final doom, on top of the economic and physical degeneration of the people. Cold air can't hold as much moisture as warm air, and so the lower temperatures brought increasing dryness. Pastures began to suffer. Snow lingered longer each winter. Churches stood empty; graveyards lay full. When Martin Frobisher landed in 1578 he met only Eskimos. The Norse colonies of Greenland had melted into the earth to be mantled by year-round snow. The pendulum of climate had swung.

Perhaps the best place on earth to grasp firsthand the alternate gripping and releasing of the land by cold is at Glacier Bay, Alaska, not far from Juneau. There the white sterility of active ice is edged by the raw brown mineral debris of land newly free from glaciers and in process of green reconquest by plants. Sixteen glaciers thrust as walls of ice into the water of today's fjords, splitting off great slabs that ride as icebergs until they melt. Camp at an inlet in the upper bay, and each outgoing tide will deliver a new array of free-form ice sculptures to your beach doorstep. The bergs come variously shaped and sized, and their color ranges from clear to opaque to blue to earth-brown (because of imbedded silt and rock). Kittiwakes and mew gulls land on them, touching down with a great sliding and skating, and arctic terns perch to ride the current, sometimes traveling for considerable distances lined up in a row.

The birds feed close to the calving glaciers, fluttering to the water to dine on crustacea that get churned to the surface by the belly flopping of the freshly calved icebergs, and by the torrential flow of meltwater. The mineral nutrients of the water are increased where glaciers reach to the sea, disgorging the rock flour they have scoured from the land and the organic debris accumulated within the ice because of the life-forms living in it and the material blown or washed onto it. Thus enriched to a veritable broth, the water supports abundant small life-forms which, in turn, sustain larger forms, even to whales. Harbor seals frequent the ice-choked upper reaches of the inlets, hauling out on pan ice to give birth to their pups in early summer, and to sun and nap. As many as sixteen hundred seals have been counted in a single half-hour observation at the head of Johns Hopkins Inlet. Twenty at a time easily come within the frame of field glasses. Tlingit Indians took seals within Glacier Bay in ancestral times before advancing ice forced them to move to new land. With the ice again now pulled back, men from Hoonah Village returned to hunt, their new, efficient weapons gravely threatening the seals until recently when a protective agreement was signed.

Ice is withdrawing from parts of Glacier Bay at a rate believed faster than anywhere else in the world. It pulls back as much as a quarter mile per year, releasing land to lie as a primal moonscape of bare silt and gravel and rock. Captain George Vancouver explored this part of the coast in 1794, but he could sail only ten miles into Glacier Bay. Ice four thousand feet thick choked the rest, walling it so spectacularly that Vancouver referred in his log to "solid mountains of ice rising perpendicularly from the water's edge." In the time that has passed since then—a period that is no time at all geologically—the ice has withdrawn from forty to seventy miles up the bay and, overall, continues to retreat each year.

This is not a steady matter, however. It depends which lobe of ice you are talking about and at what period. Part of the two centuries since Vancouver's time has been characterized by glacial advances rather than melting. John Muir camped by the Brady Glacier in 1900, nearly one hundred years after Vancouver, and he found the ice actively covering new land, not relinquishing it. An unhappy and hungry Indian man and wife living alone near the Brady Glacier reported that the stream where they had fished for salmon now lay beneath ice. The site of their people's village long since had vanished from view. Yet the fact of its existence indicates that the ice had withdrawn during the interval between Vancouver and Muir.

Born of yesterday's snows and nourished by those of today, several glaciers that feed into Glacier Bay now measure a mile wide at tidewater. Their cliff faces rise one hundred to two hundred feet above the surface, and an equal or greater extent rests unseen beneath the water. In summer the main current of the large ice tongues moves forward as much as ten feet a day, while for the glaciers as a whole movement holds to somewhere around one-third that amount. Yearly discharge amounts to from five million to nine million cubic feet of ice from a single tidewater glacier, a stupendous amount in itself, let alone multiplied by the full complement of such ice in Glacier Bay and elsewhere along the Alaskan coast.

The sight and sound of the icebergs' birth rate as one of the world's great nature experiences. Twice my husband and I have surrendered to the allure and stayed for a few days at Reid Inlet in upper Glacier Bay. One of these times we slept aboard a small cruiser anchored close to the beach. Bergs repeatedly hit us, a thundering crash separated from the eardrum by no more than the quarter-inch fiber glass of the hull. Had ice ridden into the anchor line we might have been in trouble. But none did. Instead the icy world seemed pervaded by peace. That first time we photographed black oyster catchers—long-legged, jet-black birds with bright orange chopstick bills typically held open in raucous, staccato protest at disturbance. These oyster catchers were nesting, however, and oddly silent. Their trick was to lure us by scuttling off, only to disappear behind a small chunk of ice and peer around to see if the dodge had succeeded, or if we still were looking their way.

On our second trip we joined a ranger and his wife stationed for the summer at Reid Inlet. Manya's morning chore was to walk the beach and gather remnant icebergs for the refrigerator, a food cache dug into the ground near their tent. A shoulder yoke fitted with a pair of five-gallon gasoline drums facilitated carrying more than a day's supply of ice at a time, and once we used the surplus to freeze ice cream. Packing it into the old-fashioned crank freezer, we spoke of what we were doing: utilizing the snow from storms of untold centuries past. Its fossil coldness, held by these particular glacier chunks, was about to give us the delight of ice cream to cap a summer evening at Reid Inlet.

Later we watched the calving of icebergs off the Margerie Glacier, several inlets distant from Reid. A mounting series of thundering and cracking noises would herald blocks breaking loose and, since we were in a boat a mile off the ice front, the sounds reached us a while after the action was over. Falling ice would shoot up plumes of water as it slid into the bay; then we would hear the cannon report that actually had accompanied its first break from the active front. By the time the shock wave generated by its fall had begun to subside and the sea was starting to quiet, we would hear the deep boom of the ice hitting the water.

Nowhere is the raw power of snow-turned-to-ice more apparent than at Glacier Bay—power and also the seesawing of consequences as climatic pulsations extend the reach of the ice, then withdraw it. Clusters of silvered stumps stand upright on a few beaches and erode from loose slopes. They are western hemlock trees that were overwhelmed by gravel deposits and suffocated from two thousand to seven thousand years ago, as shown by radiocarbon analysis of the wood. Glaciers were locking up a great deal of water at that time, so the ocean stood lower in relation to the land than it does today. Where there now are beaches, the trees then were growing on upland benches. The fact that they are hemlock stumps suggests a long period of forest conditions. Hemlock is not a tree to pioneer raw land. Sitka spruce does that. Hemlock comes in later.

Seemingly, gravel washed from an advancing glacier and smothered the trees during a local "little ice age," or maybe they drowned as the enormous weight of expanding glaciers depressed the land below sea level. Either way, the roots of many presently are washed by each day's tides, clear indication that they couldn't have grown recently. Such a habitat would be impossible for living trees. But at the time these hemlocks sprouted and grew, salt water stood twenty feet lower than now. When the land was lowered, the ocean encroached. Now the beaches again are rising. The general melting back of the ice is causing a rebound of an inch and a half per year at Bartlet Cove in Glacier Bay, a remarkably rapid rate of change.

This line between livability and unlivability is a fine one, and it is ruled by a very few degrees of temperature difference. It is the sort of tight margin that permitted the Greenland colony, then hastened its destruction. In Glacier Bay the wavering of the line is easy to see; the glaciers come and go rapidly. On a global basis and a great time scale, however, the evidence can be difficult to read.

Perhaps the first understanding came from the Swiss. Peasants noticed glaciers carrying boulders and concluded that large stones in their fields also might once have ridden the conveyor belt of ice. Similarly they noticed scratched and polished rock and smoothed knobs emerging from the edge of current ice, and they reasoned that these same characteristics elsewhere must be associated with glaciers of the past. By the nineteenth century, scientists led by Louis Agassiz felt sure that ice on an enormous scale previously must have blanketed much of the land. They recognized its signs on the plains of North Germany, dotted with erratic boulders from Scandinavia, and in America where rocks from Canada rested in United States wheat fields.

As the twentieth century opened, investigators realized that the presence of the great ice sheets had not been a single occurrence: On the contrary they more than once had come and gone, the cold interspersed with periods of warmth. Reindeer antlers were recovered along the Mediterranean, and woolly mammoth and arctic fox fossils came from central Europe. All gave some indication of past cold. Conversely, the remains of macaques and hippopotamuses also were found there, positive witness to previous semitropical conditions. Only sweeping changes in past climate could account for such divergent animal species within the same general region, and identification of pollen recovered from successive strata bore this out. It revealed plant species known only in cold climates interlayered with those strictly from warm climates.

Fossil evidence from the ocean bottom similarly points to climate shifts. The Eocene sea that covered Paris more than forty million years ago hosted a certain type of foraminifer, a microscopic animal, known today only in tropical waters; and even as late as Pliocene time, from twelve million to as recently as two or three million years ago, Mediterranean-type mollusks and corals lived in the North Sea. One way to trace such changes of ocean temperature through the ages is by shifts in the composition of shells. Calcium carbonate, the main constituent of many marine shells, is a chemical compound of calcium, carbon, and oxygen. The oxygen holds the key to "reading" past temperatures. More than 99 percent of it known on earth and in our atmosphere has an atomic weight of sixteen but a fraction of 1 percent has a weight of eighteen, and the ratio between these two varies with temperature. During warm periods less oxygen-18 is present, and consequently shells from such a time hold less of it than do those from a cooler period. By using this "thermometer" the rise and fall of ocean temperature can be traced.

All such techniques—from the sophistication of oxygen-isotope thermometers to the simplicity of direct observation as in Glacier Bay or the historical record as in Greenland—help to disclose what has happened, but they have little to say about why earth's climate undergoes its rhythms. Actually ice ages are rare; our human time span simply happens to fall within one. It may be nearing its close, or it may be on the verge of resurgence. Either way, the most recent ice age, the Pleistocene, began one million to two million years ago and was preceded by other ice ages. These are known to have recurred ten million, twenty-five million, three hundred million, and six hundred million years ago, and there probably were other ice ages, as well, not yet well documented. Yet, regardless of these repeated ice ages, temperate conditions have prevailed throughout most of the earth's five-billion-year history.

Various theories seek to explain why snowfall periodically increases enough to sheathe so much of the globe in ice. One possibility is associated with the recently accepted geologic concepts of drifting continents and the accompanying shifts of the polar regions. The differing distributions of land these explanations postulate would affect climate drastically, and indisputable evidence now establishes that the pattern of continents and oceans familiar on today's maps has not held constant. On the contrary, until mid-Mesozoic time about 180 million years ago, North America, Greenland, and Eurasia were joined as a single landmass; and South America, Africa, India, Australia, and Antarctica formed a second supercontinent.

Final confirming evidence of this has come from discovery of certain fossils in Antarctica during the late 1960s and early 1970s. First a New Zealand geologist, Peter Barrett, in 1967 found an unmistakable fragment from the lower jaw of an extinct amphibian something like an alligator. This one small piece, found in the Transantarctic Mountains four hundred miles from the South Pole, indicated vertebrate life that hadn't been positively known to have existed in interior Antarctica, although it had been suspected. How did such a creature get to the white continent Down Under? Swimming seems unlikely, based on knowledge of present-day amphibians which can't tolerate salt water. The question hung without answer until two years after the original find, when thirty fossils were discovered in a single day. An American team of geologists and paleontologists headed by veteran antarctic explorer David Elliott made the new finds on a snow-free sandstone cliff above the Beardmore Glacier. Among the imbedded bones were not only the remains of amphibians but also of mammal-like reptiles.

The assemblage matched those from fossil beds in South Africa, the Indian peninsula, and the Sinkiang and Shansi regions of China, all of them tropical or subtropical species. Continued investigation brought additional discoveries, permitting a confident statement concerning past land patterns. Clearly, Africa and Antarctica once were connected —widely so, not by a mere neck of land. Narrow isthmian bridges such as, for example, today's Isthmus of Panama act as zoological filters, letting certain species migrate and excluding others. This was not the situation for the ancient antarctic animals represented by the fossils. They belonged to a varied and fully developed fauna substantially matching that of Africa during the same period. The two faunas seem widely separated geographically, but actually they must have been one. The best explanation for this would be a single landmass sometime in the past.

Furthermore, at the time Africa and Antarctica were joined they weren't in their present latitudes. This is a second aspect of past land patterns that now has been realized. The earth's crust is made up of perhaps a dozen separate plates which form the sea bottom and the continents, both past and present. Driven by molten rock that wells up along rifts in the ocean floor, these plates are constantly in motion. Before discovery of the antarctic fossils, geologists had recognized similarities between rock formations in Antarctica and Africa, and they felt there must once have been a physical link but they couldn't prove it. The new paleontological evidence furnished the final thread to tie together the theory of ancient supercontinents. Past landmasses have split and drifted to their present positions. The ancient "fit" between present-day Antarctica, Africa, and India is remarkable at the one-thousand-fathom contour offshore, which quite probably is the broad connection indicated by the fossil record.

The name Gondwanaland has been given to an ancestral southern supercontinent, Laurasia to a northern supercon-

tinent. Under today's circumstances it is the northern latitudes that have the most land and the most ice (except of
course for Antarctica). We don't associate much ice with
Australia, Africa, India, or South America. Except in their
mountain ranges, these lands have no glaciers—yet they
show evidence of former heavy glaciation. Gondwanaland
was mantled by such thick ice that present-day south and
central Africa may have been more frequently glacier-
covered in the past than any other part of the world. Antarctica and Greenland are still so blanketed by ice that
geologists have little chance to see the rocks and study the
distant past.

During Gondwanaland's ice ages, tropical swamplands
dominated northern-hemisphere Laurasia, giving rise to the
great coal forests of the Carboniferous period in late Pale-
ozoic time. The North Pole lay somewhere in what now is
the northwest Pacific Ocean. How long the two supercontinents prevailed is not known, but it is apparent that
something less than two hundred million years ago their
plates began to separate. At the same time the earth's surface seems to have cooled. Ice started to accumulate at both
poles, which by then were in their present locations. When
the poles are in mid-ocean or are surrounded by low, flat
land, the entire world's climate stays fairly moderate, including the poles themselves. This had been the case earlier,
when the antarctic-African amphibians and reptiles were
alive, and when early horses were browsing the vegetation
of today's Arctic and falling prey to saber-toothed cats.

Then the pendulum swung, and sweeping changes gradually took over. The flatness of the land was followed by the
birth of mountains, and the overall average elevation of
the continents more than doubled from about one thousand
feet to twenty-six hundred feet. This contributed to cooling,
which slowly intensified over a period of tens of millions of
years. By five or six million years B.P. ("before present," the
term used for geologic time instead of B.C. or A.D.), the world
had begun to slip toward another ice age. Around one million years ago it definitely crossed the threshold, and glaciers
began another inexorable advance, this time chiefly in the

northern hemisphere, the erstwhile Laurasian lands.

Some critical new factor must have been added to bring about the advance of the ice, but what? The rising of the mountains alone seems an insufficient cause. One possibility is that the cold was ushered in by a change in the energy output of the sun, perhaps associated with the sunspot cycle. During about 90 percent of the past half billion years worldwide temperature has averaged 72° F. but it presently is 58°, and at least four times since the onset of the Pleistocene age it has fallen to 45° or less. The magnitude of such oscillations could come from a mere 8 or 9 percent change in the output of the solar furnace, and this might possibly be the outcome of either an expansion or contraction within the sun's core, which would affect the nuclear burning of its hydrogen. According to this concept, it would take only a 13 percent drop in the sun's heat to encase the earth in ice a mile thick, assuming enough water evaporated from the ocean to fall as snow; or a 30 percent rise in the sun's heat to destroy life on earth.

But perhaps the pendulum of world climate hasn't swung in response to changes in solar energy output. Perhaps it has been to differences in the amount of that energy reaching us. Water vapor hanging above the earth could reflect back as much as 90 percent of the sun's radiation, insulating and chilling the entire planet. Concentrated clouds of dust also might block the sun's rays or might provide nuclei around which vapor could condense, setting off increased rainstorms and blizzards. Volcanic eruptions seem to contribute to such dust veils. For example, most geologists credit the stupendous eruptions of Tambora in the Dutch East Indies, in 1815, and Krakatau not far distant, in 1883, with a temperature drop felt worldwide. These eruptions occurred during what already was an extremely cold period, however, and it therefore may be an overstatement to attribute the additional chilling to volcanic ash, vast as those clouds were. Tambora spewed out 240 million tons of debris lightweight enough to rise into the stratosphere; Krakatau contributed about one quarter that much. Expert opinion differs as to what effect this much dust would have on climate.

Recent support for belief in a tie between such dust veils and lowered temperature has come from deep sea-bottom cores, which indicate markedly increased volcanism during the last two million years, a period characterized by successive ice ages. On the other hand, despite this correlation, a major problem is that possible causes can't be distinguished from effects with any real certainty. Volcanism may lead to global chilling, or it may be the other way around. Climate fluctuations could contribute to volcanic eruptions by indirectly causing stresses as the ocean basins are alternately unloaded and reloaded, with water first withheld by glacier ice, then returned to the ocean because of melting. A possible response to this could take the form of explosive eruptions.

Or, regardless of the interrelationship between volcanic dust and climate, it may be that a dust zone farther above the earth than is likely from volcanic eruptions sets off global chilling. Satellite investigation now shows such zones, probably caused by meteors. Nearly one million tons of fine meteoritic debris falls to earth scarcely heeded each day, and an equal or greater amount may hang far above our planet, augmented by dust raised as large meteors collide with the moon. If the concentration of this screen were sufficient it might reduce incoming heat enough to chill the earth.

Still another explanation for periodic world temperature changes may be the variations inherent within our orbit. Known as Milankovich cycles after the Yugoslavian physicist who originated the concept, this calls for three distinct cycles that affect earth's exposure to the sun's rays. The first of these operates over a period of about 93,000 years, the second on a 21,000-year basis, and the third on a 41,000-year basis. The longest and shortest of these cycles occur because of the elliptical rather than circular nature of our orbit. If we truly circled the sun, winter and summer would be equal in length; but the actual path of the earth is elliptical with an eccentricity that varies over a 93,000-year cycle between zero and 4.3 percent. This changes the angle of segments of the earth's surface in relation to the sun. While the orbit is elliptic, as at present, we come closer to the sun each time around than is true during a circular orbit. Furthermore, at present our northern hemisphere reaches closest to the sun in January, but in 10,500 years this timing progresses again into July. This matter of varying distance from the sun constitutes the shortest Milankovich cycle.

The third cycle stems from our orbiting the sun not only erratically and at a changing distance, but also with a varying angle of tilt. Sometimes the earth's spin axis is more nearly perpendicular to the direction of the sun than at other times, and the farther off perpendicular this tilt gets the greater the seasonal differences on earth. The closer to perpendicular, the less the differences. This shift swings back and forth over a forty-thousand-year period and combines with the two other cycles to affect distribution of the sun's energy and initiate and rescind the earth's accumulation of snow.

The superimposed fluctuations couldn't have produced an ice age in the temperate world of the late Mesozoic or early Tertiary, but they may have finished ushering in the cold after mountain building and overall elevation of the continents, together with a related general cooling, already had begun in late Tertiary. A circular orbit exposes the earth to less total solar radiation each year than an elliptical orbit does, and a low degree of tilt, with minimal seasonal differences, lets snow accumulate because summers aren't warm enough to melt winters' deposits. Icefields form, and once they are present they act as heat sinks, absorbing a great portion of the calories reaching our planet from the sun. Furthermore, ice and snow bring on additional chilling by bouncing back from 60 to 98 percent of the sun's incoming rays. Vegetated land, by way of comparison, reflects only about 20 percent of incoming radiation, depending on the angle of slope, type of plants, and so on. The reflectance figure for calm water is from 5 to 10 percent.

Multiply this effect of snow and ice by the area of the earth whitened, and the magnitude of the mirroring becomes apparent. Today far less of the earth is mantled than was true at times of maximum glacial advance; yet the total area is immense, and it reflects enough incoming warmth to

Arctic people such as Eskimos are sophisticated in their adaptation to a cold, snowy environment. Their knowledge includes selecting optimal furs for clothing. For example, the texture and oiliness of wolverine fur *(above)* is ideal for ruffs because it sheds snow. Wolf fur *(below)* is next best but eventually wets and mats as snow clings to it. *(Ruth Kirk)*

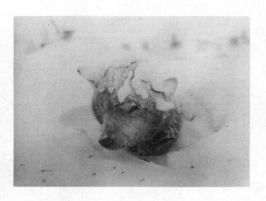

Eskimo dogs need no shelter even in winter. They simply lie curled, fluffy tails covering bare noses and bodies well insulated by thick fur. On the trail sled dogs continue pulling even when the wind-chill temperature drops to minus 125° Fahrenheit. Eager to run, they bark and leap excitedly while being hitched.
(Above: National Archives; below: Ruth Kirk)

In places, snow still eases winter travel rather than complicating it. Sled dogs pull so willingly *(above)* that "Stop" is the hardest command to teach. Even so, snowmobiles *(below)* ease tasks such as hauling firewood, checking traplines, and hunting. A breakdown or running out of gasoline, however, can be disastrous. *(Ruth Kirk)*

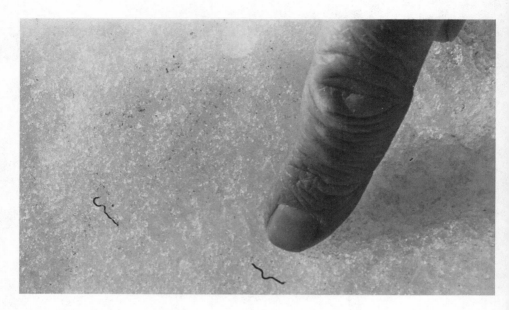

Snow worms live their entire lives
on and in snow *(above)*. Phalangids
(below right) forage snowfields and
glaciers for windborne nutrients such as
pollen and for minute organisms from
snow algae to rotifers. Winged ants
blown onto snow *(below left)*, become
prey for birds such as
rosy finches and pipits.
(Above: Ruth Kirk; below: John S. Edwards)

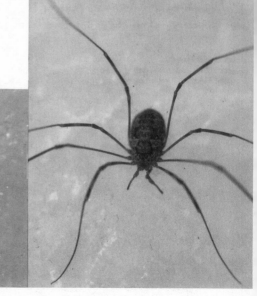

Macaques (Japanese "snow monkeys") at the primate center
in Oregon roll snowballs, then sit on them. In the wild,
young macaques make snowballs and carry them around,
but never have been observed throwing them.
(G. Gray Eaton, Oregon Regional Primate Research Center)

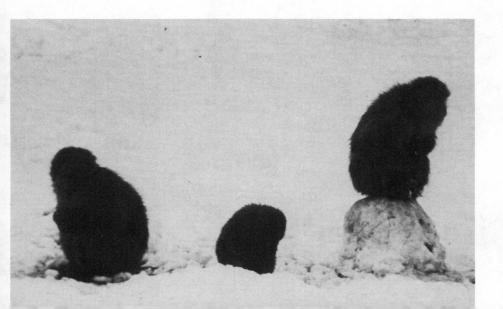

Miners stampeding to the Klondike goldfields of northern British
Columbia in 1897–98 used various means to carry food and supplies
across snowfields and glaciers. *Above:* Crossing frozen Crater Lake
en route to Dawson. *Below:* En route to Chilkoot Pass.
(Special Collections, University of Washington Libraries)

Leather booties protect the feet of Antarctic-expedition
sled dogs from the roughness of icy snow crystals,
which otherwise bloodies their paws. *(National Archives)*

Splayed hooves helped make reindeer suitable as saddle and
draft animals across Eurasia. Sami (Lapps) continue the use
as a demonstration for tourists. *(Bob and Ira Spring)*

In Norway a petroglyph (rock art) believed to be 4,000 years old indicates surprising antiquity for today's worldwide sport. Skis developed across northern Eurasia, webbed snowshoes across northern North America. *(Bob and Ira Spring)*

Early 1900s snow surveyors in Utah traveled on snowshoes when checking the depth of mountain snowpacks. They used a long, cumbersome barometer to determine elevation. *(National Archives)*

be a major force powering the world's climate machine. In 1968 satellite surveillance showed more than 7 percent of the planet covered with snow and ice in August; in late December the figure rose to 15 percent. Over the next three years northern-hemisphere photographs showed a decided increase, and by the winters of 1972–73 and 1973–74 this had risen 11 or 12 percent over what the first satellite maps had shown. The increase brought the total snow-and-ice-covered surface of the northern hemisphere to twenty-three million square miles. Furthermore, this cover had started to build nearly a month earlier in the fall and was lasting a month later into spring. So much additional bouncing away of incoming radiation potentially could bring on a rapid chilling. Will it? Nobody can say. The oscillations now have been identified but discovery alone says little about how they fit into the patterns of climate. By themselves the fluctuations of a few years mean nothing within the scheme of geologic time, and indeed overall snow cover began to diminish again in 1974–75.

For snowfall to be heavy, open water must be located where prevailing winds can pick up moisture that eventually is precipitated. A theory developed by American geologists Maurice Ewing and W. L. Donn contends that Pleistocene glaciations correlate with times that warm water is flowing from the Atlantic basin into the Arctic. This warmth prevents an ice lid from covering the Arctic Ocean and, without a lid, the water evaporates. Snowfall increases. When it becomes sufficient, continental ice sheets form. They withhold water from the ocean. This lowers sea level and in time causes the flow of Atlantic water to be cut off by a submarine ridge lying between Iceland and the Faroe Islands. Without the warm water, the Arctic Ocean freezes over. Evaporation slows, which lessens snowfall, which diminishes the ice. The glaciers begin to melt. The sea rises. The warm flow resumes. And the cycle starts anew. Considerable evidence supports this sequence of events; some contradicts it.

No matter what the reasons behind the onset of Pleistocene glaciation may have been, the nature of the ice advance itself is known fairly well. Worldwide warmth had lasted for two hundred million years, but it ended. Winter began to linger longer each year; rains changed to snow. The white blanket from one year remained to be covered the next. Ice on the high ground of Eurasia, Greenland, Iceland, and North America inched down mountain valleys, propelled by the enormous and increasing pressure of its own weight. Ice also spread in all directions from accumulation centers on plateaus and plains. When these ice caps started to wane about fourteen thousand years ago, they were covering nearly one-third of the earth's land surface. Europe lay frozen in a glacial grip from Scandinavia across the Baltic Sea into Germany, Czechoslovakia, and Poland. To the west the British Isles were covered by ice, and to the east ice reached as far as the Kola Peninsula and blanketed additional portions of the Siberian, Kamchatkan, and Asian plateaus. Offshore even the high peaks of Taiwan and Hawaii lay beneath ice.

In North America a continental glacier as much as ten thousand feet thick stretched from Alaska south to the state of Washington; another spread from Hudson Bay to Kansas and Iowa and down the east coast to New York. Boulders carried by that ice lie today in Central Park. Six million square miles were glaciated by the two North American ice caps, the Cordilleran of the west and the Laurentide of the east.

For the southern hemisphere fewer details are known, although it is clear that the last ice advance sheathed parts of Africa, New Zealand, Tasmania, the Patagonian plains, and the peaks of the Andes with glaciers. Pollen evidence is beginning to permit the tracing of climate shifts, in the southern hemisphere as well as in the northern, and carbon-14 dating is providing a skeletal time clock in radiocarbon years before the present. From this evidence it appears that the two hemispheres may have experienced slightly reversed phases of glaciation, with events in the north setting the tempo. When ice advanced there it retreated in the south, and vice versa. This occurred because of a worldwide drop in ocean level as northern-hemisphere glaciers locked up water. Additional land more than fifty miles wide ringed Antarctica, providing a greater area of support for the growing ice cap there. It crept over the land and into the sea. Then warming in the north melted the glaciers there and raised the level of the ocean. This lifted the edge of the antarctic ice and broke off great chunks. They chilled the southern ocean, and its cold slowly spread into the northern hemisphere. Falling temperatures then reversed the warm, melting trend.

Despite this phase difference, however, most geologists consider the recent glaciations of the two hemispheres as roughly contemporaneous. They point to evidence of an equal degree of glaciation in Greenland and Antarctica, and to comparable present-day recessions of ice in both the northern and the southern hemispheres. Furthermore, the rebound of land slowly freed from the weight of so much past ice shows significant similarity from one hemisphere to the other.

As might be expected, each successive glaciation largely obliterates evidence of its predecessor—and within the Pleistocene ice age alone there have been at least five major advances interspersed by warm periods. Yet even with most of the land now lying open to the sun and revegetated, the signatures of the ice endure. In the United States these vary from the magnificent sculpture of Yosemite Valley and the sharp horn peaks of the Teton Range to scratch marks and polish left on bedrock outcrops along the New England coast and as far south as Manhattan Island. Ice scoured and filled the basins of the Great Lakes but, with its great weight now gone and water impounded only behind morainal dams, the lake basins are rising about twenty inches a century. Glaciers caused the formation of Niagara Falls, shaped Cape Cod and Bunker Hill, and deposited the mineral debris that plains farmers cultivate for wheat. By melting, ice left the pothole lakes of the prairies. Half the duck population of North America now frequents them, migrating in response to inner urgings that may themselves derive from the time of the glaciers. The lengthening reach of ice drove birds before its harsh sterility, according to this reasoning. Then, as the glaciers retreated, the flocks returned to their ancestral latitudes. The pattern became established and now repeats itself on a seasonal basis.

No other single force can match glaciation in accounting for lakes, present and past. In some cases, ice scoops basins or builds morainal dams that later hold water—for example, the Great Lakes and the countless alpine tarns in which fishermen try their luck. In other cases ice itself may act as a dam. Portland, Oregon, is built on lake sediments that were swept southwestward from Idaho and Montana by flood-waters about thirty thousand years ago. A lobe of ice two thousand feet thick had blocked the Clark Fork River and created an immense lake of which today's Flathead Lake in Montana is the only existing remnant. Pressure against the dam eventually broke it and five hundred cubic miles of water raged southwestward, carving the channeled scab-lands of eastern Washington, rafting huge boulders on ice-bergs, and turning the Willamette Valley—three hundred miles away—into a lake over one hundred feet deep. In somewhat the same way as the ancient Clark Fork ice dam, the Taku Glacier at Juneau's doorstep currently is expected to block an inlet and impound water that will reach deep into British Columbia. If the Taku continues its present rate of advance this will happen within a decade.

Nearly five million square miles of North America were glaciated during the last ice age; so were more than two million square miles of Eurasia. In Antarctica, South America, Africa, India, and Australia perhaps another five million square miles lay beneath ice. Most of the vast weight of these glaciers now has been lifted, and the earth's crust is rising at a rate perceivable within a single human lifetime. North of the Baltic Sea at a narrow point of the Gulf of Bothnia the land is coming up almost half an inch a year; reed banks there turn into meadows, and harbors continu-ally silt in and need dredging. In Canada the floor of Hud-son Bay has rebounded about nine hundred feet since the retreat of the Laurentide ice sheet, perhaps eight thousand years ago. It continues to rise and is expected to lift another eight hundred feet, which may drain its water. National Park Service personnel at Glacier Bay have to wait for high tide to get their boats in and out of the anchorage near their houses because of the crucial sea-bottom rise of that particular cove in the years since the headquarters area was designed and built in the early 1960s. On the west coast of Washington at the Ozette archaeological site the oldest evidence of sea-mammal hunters comes not from the deepest deposits but from the highest, ancient beach terraces now forty-five feet above the lap of waves. At Alaska's Cape Krusenstern, a horizontal succession of old beaches yields a sequence of artifacts ranging in time from recent habitation by Eskimos to the cultures of progressively earlier peoples. Recent items lie close to the present beach; older items are on the ancient beaches parallel to today's water but no longer washed by its tides.

Changes of sea level in relation to the land are reported from all over the world. They are not localized occurrences but global. Their mechanisms are complex and only partly understood, although it seems likely that the level of the water is a direct response to the transfer of volume and weight from ocean to land and back to ocean again as water is held by ice, then released. The most recent worldwide glacial maximum accounts for a theoretical ocean-level drop of about 360 feet. By 6,000 years ago melting had brought the level up to essentially what it is today. Then 3,000 years ago a climate warmer than today's raised the ocean nearly seven feet above what it now is, and sea ice in the Arctic may have melted each summer. If all the ice presently remaining in the world should melt, a rise of 150 to 200 feet above the current ocean level could be expected. This would be offset partly, however, by the isostatic rebound of the land freed from the ice's weight.

A variety of additional factors also may contribute to the repeated submergence and reemergence of land. Sediment deposited from the continents onto the ocean floor displaces water and raises sea level, although nobody knows for sure the magnitude of this in the past. Estimates are that if present-day land were torn away and carried into the ocean, water level would rise a possible eight hundred feet, but so far as is known nothing even approaching such a wholesale deposition ever has occurred. Besides, sediment loads on the sea floor also trigger crustal displacements along fault zones, complicating the task of tracing their specific effects. Also, cooling or heating of the entire ocean must affect water level, although a full ten degrees either way are calculated to bring about only a seven- or eight-foot rise or fall.

The fairly newly discovered bombardment of our planet by protons from the sun may furnish raw material for new water, which at least theoretically could raise the level of the ocean. The effects of this aren't really known—some scientists aren't even sure the phenomenon exists—but it may be that the incoming protons, which are hydrogen nuclei, combine with oxygen in the outer atmosphere to form water which precipitates to earth. This would increase the global supply of water, previously believed an unchangeable and eternally cycling amount. What the potential increase from this source may be isn't yet speculated upon.

Contractions of the earth are another possible cause of raised sea level, because of reduction in the size of the ocean basins. Conversely, a general expansion would stretch the basins and lower the water level, and the spreading of ocean-floor plates now is known to occur. This apparently isn't affecting the size of the world, however, because as new magma is added to the ocean plates, continental plates undergo compression and mountain building, and they also have their edges driven back down into the mantle.

Of the various explanations for major ocean-level shifts, the tying up of water in glacial ice remains the most plausible, and it alone is enough to account for radical coastline changes known to have taken place. For instance, within geologically recent years England has been joined to the mainland, Sicily connected with Italy, New Guinea with Australia, Sakhalin Island and the Korean peninsula to the Asian continent, Siberia to Alaska. The Bering Strait, even now only about fifty miles across, once was a sagebrush-covered steppe. The sea bottom beneath today's northern

Bering Sea and almost all of the Chukchi Sea is one of the flattest, smoothest expanses known anywhere. It stretches as a drowned plain from the eastern Aleutians to Cape Navarin south of the Chukchi Peninsula, through Bering Strait, and along the arctic coast from the Mackenzie River delta, past Wrangell Island and the New Siberian Islands to the Lena River and beyond. Known as Beringia, this plain is about nine hundred miles wide from north to south, which is considerably greater than the width of Alaska along the Canada border.

Evidently Beringia has alternately stood above the ocean and been flooded. Once it belonged to Laurasia, and turtles and alligators lolled at ease in its marshy lakes and lemurs clambered through its forests. The flora and fauna of today's eastern Eurasia and North America were then essentially one, as was true for the fauna of Antarctica and Africa. Subsequently Laurasia broke apart on a line about along the present-day Lena River, then rejoined. Since then Beringia has at times been awash and at times above the water. By the onset of the most recent glacial advance its land repeatedly had emerged and been inundated and was about to make still another appearance. This most recent ice advance, known as the Wisconsin in America and the Würm in Europe, began somewhere around 100,000 to 80,000 B.P. It reached a maximum about 40,000 B.P. and by 14,000 years ago started to wane.

Sometime during this period, while glaciers were holding water in their icy fastness and Beringia was above water, man arrived in America. He hunted his way across from Asia seeking woolly mammoths, giant bison, elk, moose, caribou, and other animals which were grazing and browsing the broad steppe that joined what now are two separate continents. Just when the first Americans arrived, exactly where they came from, what they were like at the time, and how they populated the New World are questions only beginning to be understood. Little more than hints of man during these millennia are ever likely to be glimpsed, for the early arrivals must have been few and have lived in scattered family groups.

Nonetheless, camps and hunting sites belonging to these ancient newcomers have been found in every nation of the New World. Fragments of human skulls more than ten thousand years old have come from Marmes Rock Shelter in southeastern Washington, the oldest actual new-world human remains discovered so far in a context that can be positively dated not only by radiocarbon but also geologically and by cultural, faunal, and floral association. Artifacts, animal kill-sites, and fire hearths of equal and greater age have been found at various locations scattered all the way to the tip of South America, with hunting camps in the highlands of Peru dated at nearly 22,000 B.P. and cultural deposits in caves along the Strait of Magellan at about 8,000 B.P. North American dates, as determined by radiocarbon means and geologic context, are as old as 40,000 B.P., and archaeologists believe there must be still older sites they haven't found yet.

Some archaeologists speculate that man may have come to America as long ago as 70,000 years. Beringia is known to have stood above sea level sometime before 35,000 B.P. (probably repeatedly so), again between 25,000 and 18,000 B.P., perhaps at 15,000 B.P., and seemingly again as recently as 12,000 to 11,000 B.P., although this last still is somewhat conjectural. While the bridge of land existed, man could walk from Asia to America, gradually extending his territory; and while Beringia was submerged he may well have paddled across as Eskimos do today, or walked across on winter sea ice. On clear days at present Cape Prince of Wales, Alaska, is visible from Cape Dezhnev, Siberia. Old World antecedents for each wave of migrants are not known well partly because investigation of sites in Asia hasn't been widely reported to Western archaeologists. Even aside from political and language barriers, the press run of Mongolian and Siberian journals is small, and distribution is limited. A great deal more comparative study must be done before the full picture emerges, but general similarities are recognized between artifacts from Onion Portage, Alaska, and those from Lake Baikal, Siberia, and scattered sites in Japan, Mongolia, and Kamchatka. Clearly, man came to America from Asia, arriving in the Arctic and then extending his range southward.

By extending territory only four miles a year, man could have moved a thousand miles farther south every two or three centuries, according to Edward Deevy of Yale University. He assumes a band of thirty people—five families each with two parents, two children, and two grandparents—starting along the ice-free corridor that at least during Pleistocene time is known to have fingered between the Cordilleran and Laurentide ice caps. Using population expansion factors of 1.2 and 1.4 for each generation, the original thirty people would have numbered between eight thousand and twelve thousand at the end of the first five centuries. They probably would have separated into around four hundred distinct bands, based on the land's carrying capacity under the conditions of the time. This works out to little more than one person per square mile and a group size of about twenty-five persons needed for cooperative subsistence and sound genetics.

On the basis of these figures man could have reached into Central America in the first five hundred to one thousand years after crossing from Asia, and another few hundred years would have taken him to the tip of South America. Many archaeologists don't agree with such rapid expansion, however. They feel a factor of only .5 per generation is more reasonable than 1.2 to 1.4. This in itself would more than double the time needed for getting all the way down the South American continent. Furthermore, since people moved not by prearranged plan but according to subsistence needs, they were not impelled to stay on the march. On the contrary, whatever their maximum rate of expansion may have been, they probably didn't achieve it. There wasn't reason enough to do so. On the contrary, learning to cope with changing environments and resources would be a slow process undertaken only as absolutely necessary. Nobody would move simply for the sake of moving.

To know about this, rather than speculate, however, an immense amount of additional evidence is needed—and, sadly, the chances of getting it aren't altogether promising.

The number of sites never can have been many, and preservation within them seldom is more than sketchy. Worse still, evidence of prehistoric man is being destroyed rapidly around the world both by the deliberate digging of amateur collectors and the legitimate planting of fields and construction of highways and dams. The river valleys and mountain passes most useful to early man remain the valleys and passes useful to man today.

Lowered ocean level eased man's arrival in America, and today's level determines the location of port cities such as New York, London, and Tokyo. Yet if the glaciers now present were to melt, water would rise three hundred feet, and unless the land also rose, barnacles would grow on the Empire State Building and sea anemones would wave from Buckingham Palace and the Ginza. Practically every culture has a myth chronicling a flood of such magnitude. For instance, Panamint Indians whose home is in Death Valley, California—by some reckoned the hottest desert in the world today—trace their ancestry to escapees from a flood that filled the valley hundreds of feet deep, leaving only the surrounding mountain rim as refuge. Their tale has a basis in fact. The great Pleistocene ice sheets of the north had a pluvial counterpart beyond their southern reaches, and lakes flooded much of what today is the arid southwest of the United States.

New evidence linked to glaciation even lends possible credence to the ancient story of Atlantis in the Mediterranean region. University of Miami paleoclimatologist Cesare Emiliani and co-workers from various disciplines have studied a series of cores taken from the Gulf of Mexico not far from the Mississippi River mouth. Isotope concentrations in fossil shells from a core depth of about five hundred feet indicate a rather sudden drop in water temperature there at the time the animals within the shells were alive. Radiocarbon counting puts a date of 9600 B.C. on the chilling, and this coincides with Plato's accounts of the disappearance of Atlantis. The Laurentide ice sheet seems to have given a last surge forward, then rapidly melted along its southern edge. Runoff caused a great deal of flooding inland and raised the ocean level by several inches a year. Multiply this by similar occurrences elsewhere, and low-lying land unquestionably would have vanished from sight. Traditions from Atlantis and Noah's ark to floods in Death Valley have a basis in fact.

Our globe still wears an ample glacier mantle. Whether it now is growing and will lower the ocean level, or is shrinking and will substantially raise the ocean, isn't really known. Greenland alone is weighted by ice two miles thick, its bedrock depressed like a saucer. Only the highest peaks are exposed. Snowfall there yields a yearly equivalent of 107 cubic miles of water, and even more melts than falls. Antarctic ice exceeds Greenland's volume by about seven times, a total calculated as seven million cubic miles. This mentally staggering quantity seems to be neither increasing nor decreasing greatly, although the statement is based on too little evidence to be more than tentative. Assuming the figures to be reasonable, the combined antarctic and Greenland ice represents 99 percent of present-day glacial ice which, in turn, amounts to no more than one-third the volume of ice at the height of the most recent ice age. No other solid component of the earth's surface is as widely distributed as ice. Glaciers cover more than 10 percent of the land surface: a little over six million square miles, which is about the same as the amount of land that is farmed, or of the tropical rain forest belt. Sea ice is distributed over more than twice the area of terrestrial ice, and if icebergs broken from glaciers are added to this figure the total for floating ice becomes more than twenty-eight million square miles—23 percent of the ocean's surface and 14 percent of the entire globe's surface.

Is the ice now dwindling? Or is it building?

"Maybe so," answers both questions. Studies are too recently begun and lines of evidence too conflicting for anybody to feel confident one way or the other. To know is to heed only one set of data. What's more, man himself probably is a contributing factor, with no means yet of knowing which way.

The broad outlines of what we know suggest that the earth is vastly cooler now than when it first formed and, from the standpoint of life, that its temperature has been oscillating between moderately warm and intensely cold for the last one or two million years. We may be finishing one of the moderately warm periods now. Indications of this have been the recent thickening of some mountain glaciers, first noticed on the Coleman Glacier of Mount Baker, Washington, in 1948 and on Mount Rainier's Nisqually Glacier in 1951. Prior to that, world climate had seemed for fifty years or so to have been about as it was during the time of the great Viking expeditions: warmer than average.

The most recent—now possibly ending—period of warmth and the earlier warmth that favored the Norse were separated by hundreds of years of biting cold. Napoleon's problems in Russia were compounded by the bitter cold of winters during that period. A third of a century before his time, the British forces occupying New York City also had cursed the problems of winter. The American revolutionaries were worse off, hunkered around their campfires at Morristown, New Jersey, but the British suffered the loss of their fleet. It was locked useless for weeks at a stretch by the freezing of much of the New York harbor, although cannon could be hauled across the ice from Staten Island to Manhattan.

During those times all that anybody could do was to get on with the tasks at hand despite the cold, or give them up because of it. Climate, which is long-range weather, swung on a pendulum beyond man's reach. Soon this may not be true. We may be able to alter the weather, and quite possibly the climate. Some of this is intentional. Experimental cloud seeding was begun in 1946 by Vincent Schaefer, now head of the Atmospheric Science Research Center at New York State University, and Irving Langmuir, a Nobel laureate. Their work established that substances such as silver iodide can be dispersed into the atmosphere to act as nuclei for ice crystals which ultimately grow into snowflakes or raindrops. These introduced substances may even function more effectively than naturally occurring nuclei, which often are sparse.

Experimentation continues. In the 1970s, in the United States alone, the Bureau of Reclamation, the National Oceanic and Atmospheric Administration, the Department of Transport, the Forest Service, the Federal Aviation Agency, and various military agencies all have worked at weather modification. Goals vary. So do results. By way of illustration, Project Skywater recently seeded clouds in western Washington and in the immediately ensuing winters all previous world records for snowfall were broken at Mount Rainier. Communities on both sides of the Cascade Range found their snow removal budgets strained to the breaking point. It can't really be said this stemmed from the cloud seeding, however. The winter may have been a naturally snowy one.

Skywater was a Bureau of Reclamation undertaking. From 1970 to 1975 another Bureau of Reclamation project seeded clouds in the San Juan Mountains of Colorado for the purpose of "winter orographic snowpack augmentation," or *WOSA*, to use official jargon. The WOSA was expected to produce 2.3 million acre feet of water, or *maf*, per year within the Upper Colorado Basin and a 1.2 maf in the surrounding country. Yet the same National Science Foundation report that gives these figures mentions five pages later that as little as two maf of increased water per year would increase the sediment load of the Colorado River a probable 12.5 million tons per year. This can't help but hasten the filling in of Colorado River reservoirs. It also would raise the salt concentration of the river by 700,000 tons. WOSA melt can't be expected to dilute maf's increased salinity because a greater supply of water will mean greater evaporation and greater agricultural and industrial withdrawal. Treaty obligations to pass on water of usable quality to Mexico almost surely will be as difficult as ever to honor.

Such examples of man's intentional efforts to increase precipitation are legion. Even more numerous, and potentially more far reaching, are his unintentional effects. Add lead particles to the atmosphere, as our automobile exhausts and industrial chimneys do by the ton, and ice-crystal formation is stimulated just as effectively as by the dispersal of silver iodide particles. Add too much lead, and clouds may even be prevented from releasing snow and rain. The ice crystals form but stay too small to gather vapor droplets, grow, and fall to earth. They don't collide with one another, collecting and merging into a size that can overcome gravity. Laboratory tests from around the world repeatedly have produced such overseeding, although precisely what is happening currently on a global scale isn't known. One likelihood is increased cloudiness and suppressed precipitation as ice particles too tiny to fall ride the winds above the earth. Or, if they meet an infusion of warm moist air, the effect may be devastating blizzards and rainstorms. Artificial cirrus cover similar to that produced by automobile and industrial pollution also results from lingering—and increasing—jet airplane contrails. Their vapor lasts in the upper atmosphere an estimated eighteen months.

The consequences of what we are doing aren't known well but the statistics of the actions themselves are beginning to come in. Human-caused pollutants presently constitute a frightening 326 million tons of particulate matter spewed into the atmosphere each year. (A single copper smelter in Tacoma, Washington, belches out 22.2 tons of sulfur particulates *per hour*.) Dust rising from earth accounts for another 100 million to 300 million tons per year. This comes from the mechanical tilling of ever more acreage as demanded by increasing human population, plus recent wholesale drought conditions and the man-induced advance of deserts. Smoke from agricultural and slash burning adds a further 45 million to 65 million tons of matter. What are the implications? One may be to screen us from the warmth of the sun's rays. If the natural swing toward a colder climate, which many experts believe they detect, is actually the case this human-caused atmospheric veil may hasten and intensify the recurrence of an ice age.

On the other hand, global chilling may be offset by other

factors. Gases from our aerosol spray cans may diminish the protective ozone layer high in the stratosphere and condemn us to too much ultraviolet radiation. Billions of pounds of aerosols per year have gone into the air from America alone as householders have sought odor-free bodies, clean ovens, and shiny paint jobs on cars. The real result of all the spraying may be a slow incineration of earth, or at the very least an alarming increase in skin cancer. But it also may not. Nobody is sure.

An effect that can be stated definitely is a marked warming above cities. Stone and concrete and asphalt store up more heat than vegetation does. As a result urban temperatures range as much as four degrees Fahrenheit higher than those surrounding rural areas in summer and two degrees higher in winter. Plumes of hot air reach for half a mile above some cities. A survey in the United States shows an increased rainfall of from 9 to 27 percent in nine cities, seemingly as a result of such heat islands. This is because the heated air rising from cities reaches altitudes high enough to condense its moisture. Rain now seems to be more likely than snow over cities because of this new, artificial warmth. Hail definitely has increased. For example, Houston, Texas, experiences more than four times as many hailstorms as it previously did, largely because of this one factor. La Porte, Indiana, in 1965 showed 35 percent more thunderstorms and nearly 250 percent more days of hail, compared to previous long-term weather records. Over the last forty years La Porte's precipitation has risen 30 to 40 percent, which parallels the rising curve of nearby steel and iron production in Chicago and Gary, Indiana. Conversely—and typical of interlocking and contradicting bits of evidence—the recent five-year drought in northeastern United States also is attributed to urban heat barriers. Rain clouds that otherwise would have drenched the region were forced to flow around the sprawling megalopolis.

Another factor possibly acting to warm the world is the renowned greenhouse effect produced by a floating blanket of carbon dioxide. Our air now holds 10 percent more carbon dioxide than a century ago, most of it added since

World War II because of accelerated fossil fuel consumption. This source currently feeds twelve billion tons of carbon dioxide into the atmosphere every year, which may keep heat from continuing to escape normally from the earth. Incoming shortwave solar radiation passes easily through the natural atmospheric veil and can keep on doing so even with the increased carbon dioxide concentration; this is similar to the way heat comes through the glass of a greenhouse or the windshield of a car. The earth sends the heat energy that it doesn't absorb back out as long-wave infrared radiation, and normally this is lost to the atmosphere. Some observers believe that a change now is underway. Infrared waves are absorbed by carbon dioxide, not transmitted. Consequently excess carbon dioxide may hold heat against the earth rather than let it escape, just as glass holds heat within a greenhouse. Some feel that they see the beginnings of this now.

If true, many Cassandras of climate predict that the heat will melt the world's ice. But whether they are right, or their counterparts who argue for an imminent return of the glaciers, isn't ours to know—at least yet. Whichever way it goes, the fragile balances of the high latitudes will show the first drastic effects, as experienced long ago by the Vikings in Greenland. The vigils kept in the white nether reaches of the earth by today's scientists beyond question carry great import. We don't yet know just what it is, but data are the raw ingredients of understanding.

CHAPTER

3

TO THE POLES

THE GREAT SNOW REALMS OF THE ARCTIC AND ANTARCTIC
have drawn men to them ever since early whispers of their
existence began to reach temperature-zone adventurers and
scholars. Traveling across the frozen wastes held horrors
beyond ordinary imagining until recent decades, yet men
went and persevered. Apsley Cherry-Garrard on Scott's
second expedition to the Antarctic, 1910 to 1913, wrote:
"They talk of the heroism of dying—and they little know—
it would be so easy to die, a dose of morphic, a friendly
crevasse, and blissful sleep. The trouble is to go on."

But "on" they went so long as breath remained; scores of
men prompted to stay the course even when strength and
reasonable hope were gone. Why? For Scott's men, Cherry-
Garrard put it simply: "We traveled for science . . . in or-

der that the world may have a little more knowledge, that it may build on what it knows instead of what it thinks. . . . Whilst we knew what we had suffered and risked better than anyone else, we also knew that science takes no account of such things; that a man is no better for having made the worst journey in the world; and that whether he returns alive or drops by the way will be all the same a hundred years hence if his records and specimens come safely to hand."

Few men from those heroic days of polar exploration remain to talk with today, to ask what it was like and what drew them to the snowy nether reaches of the world. But one lived quietly on Salt Spring Island in British Columbia until his death in 1975: Sir Charles Wright. Through introduction by a mutual friend I sat by his hearth one drizzly day that turned out to be only a few months before his death. Sir Charles was retired—once from the British Royal Naval Scientific Service, which he headed; a second time from the Marine Physical Laboratory of Scripps Institution of Oceanography; then from the Pacific Naval Laboratory; and finally from the University of British Columbia.

He had been a Canadian scholarship student studying physics at Cambridge when he heard Captain Robert Scott lecture on his first antarctic expedition, 1902 to 1903. The second expedition was then forming; Wright decided to apply for it and was accepted as one of ten men comprising the scientific staff. No previous commander had given such heed to the scientific opportunities afforded by the Antarctic but, as Sir Charles put it, "Scott was a fine scientist lost to the Royal Navy." He had tried to explain the origin and feeding of the great ice formations he charted, such as the Ross Barrier and other ice shelves. He had been the first to advance the idea that sometime in the past a lowered temperature had brought on a lessening of precipitation, which in turn could account for the partial deglaciation of the great southern continent.

Financing for Scott's second expedition came from the British government, the Royal Societies, and also from the public. Nobody yet had reached the South Pole, and hope ran high that this expedition would be the first to get there. Because of this public appeal, reaching the Pole ranked as Scott's own primary goal, but, for him, scientific observation and the collection of specimens came second only because of the limited time and energy available, not because of less interest. Meteorologists, physiographers, geologists, a biologist, an oceanographer, an ornithologist, even a helminthologist (an expert on one type of worm) belonged to the scientific staff. Wright was a physicist who became a glaciologist, one of the world's first, for study of the physics of ice had barely begun in the European Alps at the time, and little was known of the nature of polar ice caps. Sir Charles dropped into the job. "I didn't want to be stuck with thermometers as an assistant meteorologist," he told me. "That was my fear. I wanted to get about and see the country." So he suggested glaciology to Scott. He had no special training in the field; virtually none was to be had. But he was a scholar, and observation and study soon made him knowledgeable.

In Antarctica each day's travel demanded enormous stamina. How much so varied according to peculiarities of the snow surface. Where wind had done no packing and polishing, individual grains of snow tended to be coarse and had about the effect of sand so far as the sleds were concerned. In a similar way, if temperatures were too low for the friction of the runners to melt the points of the crystals and smooth the way, progress came only as grains moved against one another, creating a dreadful drag rather than permitting a glide. At other temperatures, warm enough to melt the snow, water would freeze onto the runners in hard lumps that interfered terribly with travel. Up-ending the sled and knocking them off was the only way to get rid of them; that meant stopping, then starting up again with desperate, energy-sapping jerks to break the sled loose and get it sliding. Most of the time Scott's party hauled the sleds themselves, hitched to them by canvas bands worn across the abdomen. They pulled in pairs, the weight of their loads as much as 250 pounds per man. Since each tug of the sled pressed intestines against spine, everybody was keenly aware of variations in the snow surface.

"How difficult the going was all depended on the surface," Sir Charles summed up the matter. "One was good for the ponies but bad for the sled runners; another good for the motor sledges we were experimenting with, but bad for both ponies and men. It depended mostly on temperature and the hardness of the snow surface." At forty below zero a man had to pull about three times as hard to move a sled as when the temperature had risen to zero. Dogs apparently had an advantage with their four legs. They aren't in step with each other as men are, with only two legs; so the dogs' pull is more even.

Scott wanted to test various types of transport. Dogsled travel was an established means of crossing snow and ice, but Scott's men weren't really familiar with it. The climate of Britain gives little opportunity to grow up with snow. Partly because of this, although Scott wanted to use dogs, he didn't plan on them for the longest hauls. The problems of carrying enough dog food when away from the coast seemed insurmountable. The men themselves would need all possible space aboard the sleds for their own supplies, and there would be no seals or penguins or other animal life to hunt while inland. Consequently, so Scott reasoned, there was no way to rely heavily on dogs. Scandinavian explorers, on the other hand, including Roald Amundsen who beat Scott to the Pole, felt perfectly comfortable with dogs. They simply let the bitches whelp and regarded the dogs as self-renewing. The flesh of those that were worn out could be fed to the rest, and at times they even provided food for the men themselves. British attitudes precluded this.

Totally new to the ice were the three motor Caterpillars that Scott had ordered specially built for the expedition. They traveled forty miles, then quit, failures from the outset except in the sense that no intelligently designed experimental model is a failure so long as it provides information that may lead to improvement and ultimate success. Scott's engines overheated, and the chain drives refused to work right. Nonetheless the motor sledges became the forerunners of the tanks used in France during World War I and the ancestors of current oversnow vehicles from individual snowmobiles to the large passenger models that take tourists onto the Columbia Icefield in Jasper Park, Alberta, and into the winter heartland of Yellowstone National Park, Wyoming.

Scott's main transportation hope rested on ponies and mules, neither of which actually worked out very well. Ponies had been used in Antarctica by Ernest Shackleton between the time of Scott's two trips. Shackleton had heard of their staunch indifference to snow and cold during the Russo-Japanese War and decided to try them on the antarctic ice. Results were so good that Scott too planned on ponies, taking aboard nineteen as he sailed from New Zealand. Most were Manchurian animals, the rest Siberian. Their rations were oats, oil-cake, a little hay, chaff, and bran to be cooked into a hot mush. They needed more food but nothing could be done. Where in all of Antarctica could it come from? There is no local vegetation to draw on.

In temperament the ponies were a mixed lot—"some slow, some fast, some kind, and one that was dreadful," according to Sir Charles. "You had to throw him to get him into harness." The process took half an hour, so men with that pony, named Christopher, gave up stopping for lunch on their day's march. It left them hungry and tired but avoided the need to get Christopher going again after a stop. The ponies were an advantage in being able to pull as much as a half ton, under the right conditions, but overall they spelled trouble. The best snow surface for men pulling sleds was about −16° F., but for the ponies safe travel was at night when temperatures usually were colder than that. They could rest and sleep only during the day when the sun was high enough to dry their coats; then would work at night no matter how miserably cold. For them, marches began at midnight.

The mules Scott had along proved even more disappointing than the ponies. Snow wasn't a problem, as had been feared. Round snowshoes strapped to their hooves satisfactorily prevented sinking in. The mules moved faster than the ponies, pulled heavier loads, and kept together better. But they wouldn't eat. Oats, compressed fodder, and oilcake were their intended food. What they willingly swallowed instead were tobacco, the soft straw the men used as linings in their skin boots, dog biscuit when it could be pilfered but not when it was offered, rope, and harnesses. "Probably it was our fault," Sir Charles said. "We should have taught them to eat snow. We were told they were Tibetan mules but actually they came from India and probably they'd been given water there and didn't know how to eat snow. Maybe the food wasn't right either, but without enough water it's no wonder they wouldn't eat well."

Wright became navigator on the long marches, a job of awesome responsibility. At times falling or blowing snow made it impossible to see for more than thirty yards, yet travel had to continue. Peculiar lights played tricks. Tracks looked raised instead of depressed, and although the mind said not to bother, the feet kept stepping higher than necessary to avoid tripping. It was an annoying and tiring situation. Similarly, what seemed like a herd of cattle ahead, but on the basis of likelihood had to be discounted as probably dogs, would actually prove to be pony droppings. You might see what looked like the hut ahead, only to find it no more than a discarded biscuit tin. Sastrugi, the ridges and pinnacles of snow sculpted by wind, sometimes looked like impassable barriers instead of foot-high nuisances, and whoever was determining route needed to be on guard. The sastrugi also helped, however. Since they form as wind-driven waves of snow, they lie parallel to one another and dependably indicate direction. By traveling down them or crossing at a steady angle Wright could know he was staying on course. Wind itself served in much the same way. Feel it constantly at your back or stinging one cheek, and you knew the course.

Sir Charles used a compass, an ordinary one with a sliding weight on the needle to correct for the angle of depression caused by proximity to the magnetic pole. A liquid compass is useless in that location. "It usually leaks and even if it doesn't, it takes so long to settle down that you do better with an ordinary compass. I'd go along a certain distance and think it time to have another look, then pick up the march again. Lots of things let you stay on the right line if you keep your wits, as you must when you have the unpleasant job of navigating.

"Maybe it's the sun you go by, or the moon, seeing them relative to the direction you want to go. Even under the worst conditions there are little things you can use—a bit of light in the sky over here, which you don't really look at but can use, or ripples in the snow at right angles to the wind of the last blizzard. You can feel ripples underfoot even if you can't see them. Maybe there's no wind and no drift and you can't see the horizon: then you don't know where you're looking. You think you're focused at infinity, but actually you're looking at the tip of your ski; you seem to see one of the cairns from the outgoing trip, but it proves to be your ski. A little bit of black really hits you in all that white.

"The navigator has to see all that's to be seen. It's a good idea to use one eye for walking along and save the other for your job. There's no sense in getting snowblind. With a certain amount of intelligence it can be avoided. The thing to do if you feel it coming on is to avoid the conditions that produced it. Wilson got it on the Beardmore Glacier, yet there he was at midnight out sketching."

Probably such tenacious devotion to art was foolish, but the sketches and watercolors by Bill Wilson, expedition ornithologist and director of research, carry the coldness and awful remoteness of the land more fully than any other rendition. Photographs from today's larger antarctic expeditions, with their advanced technology, perhaps capture the landscape more fully. But Wilson's hand recorded the vision of his eye and heart with a selective force beyond the capacity of lens and film.

From Scott's base camp at Hut Point, the Pole lay nearly nine hundred miles away. Sixteen men, including Wilson,

Cherry-Garrard, and Sir Charles, set out for it with ten ponies, twenty-three dogs, and thirteen sleds. The plan called for establishing food caches on the outward journey so that the men could supply themselves on their return journey. All would travel to within striking distance of the Pole; then a small team to be picked by Scott would press on while the others turned back. The motor sledges started on the trek, too, but had to be abandoned. The ponies were expected to give out, whereupon they would be sacrificed to feed the dogs, each providing food for an additional four days. When the dog food thus augmented ran short, both dog teams would turn back with four men. The other twelve would continue, man-hauling the sleds loaded with food, oil for the primus stove, tents, and spare sled runners.

New Year's Day, 1912, found the party of twelve 180 miles from the Pole, and Scott that day decided on five men to make the final push: Scott himself, age 43, Wilson, 39, Edgar Evans, 37, Titus Oates, 32, and H. R. Bowers, 28. On January 4 the support party traveled a few miles with the polar team; then, all seeming in order, they turned back. "A last handshake and good-bye, I think we all felt it very much," Cherry-Garrard wrote. "They wished us a speedy return and safe, and then they moved off. We gave them three cheers, and watched them for a while until we began to feel the cold. Then we turned and started for home. We soon lost sight of each other."

As it turned out the loss was forever, the good-bye final. Scott's diary, found with his body, told of the journey. On January 16, twelve days after the others had turned back, the five men of the polar party found ski and dogsled tracks and with sinking hearts estimated them to be about three weeks old. They knew that Roald Amundsen also was pushing toward the Pole. The tracks had to be his, and they meant the race already was lost. "The Norwegians have forestalled us and are the first at the Pole," Scott wrote. "It is a terrible disappointment and I am very sorry for my companions. Many thoughts come and much discussion we have had. Tomorrow we must march on to the Pole and then hasten home with all the speed we can compass. All the daydreams must go; it will be a wearisome return. . . . Great God! this is an awful place."

The five men had been underway for two and one-half months at the time Scott wrote. Barely more than two months later all lay dead, three of them only eleven short miles from a cache with ample food and oil to have sustained them. Evans fell into coma and died in mid-February, following a fall. Oates perished a month later. His feet were hopelessly frostbitten, and he asked to be left rather than imperil the others. They refused. So Oates stepped out of the tent saying, "Well, I'm just going outside and I may be some time," and he hobbled off into the blizzard. His body never was found. Searchers led by Sir Charles found the tent of Scott, Wilson, and Bowers with their bodies inside, the records of their observations kept to one week before they made that last camp and found themselves pinned down by blizzards and creeping weakness. With the men were a few last personal letters, several rolls of film, and thirty-five pounds of geologic specimens including fossil impressions of plants found in a coal seam. There also was a note for King Haakon from Amundsen, which had been left at the Pole with the expectation that Scott would carry its message to the world should the victorious Norwegians themselves perish on the return trip.

At the Pole Scott had written: "He [Amundsen] has beaten us. . . . We have done what we came for all the same and as our programme was laid out." So they had. They had completed two major geologic expeditions aside from the work accomplished on this last tragic journey, and there had been a winter ornithological expedition with temperatures as low as —70° F. even at high noon. The purpose of this undertaking was to collect eggs from an emperor penguin rookery, since emperors at the time were believed to hold a key to the puzzle of evolution and fetal chicks were considered the most desirable stage for study.

Scott and his men had accomplished their "programme." But Amundsen had beaten them to the Pole. He started from a base sixty miles closer to the goal than Scott had chosen. Furthermore, reliance on dogs meant he could start earlier in the season than Scott could, using ponies and hoping with them to avoid the terrible man-hauling that actually became necessary. Also, Amundsen's men had been skiers from childhood, and they started their last push for the Pole in top condition, rather than already half-exhausted from man-hauling sleds. They rode their dogsleds for the first one hundred miles of their journey and for the next three hundred were towed on skis behind them. They weren't strangers to snow. They knew how to make it count as ally, not enemy. And they happened to pick a much better route for weather than Scott's which lay over the Beardmore Glacier, the world's largest glacier and now recognized as squarely in a blizzard zone.

On the outward journey from England, Scott had received a telegram forewarning him of the competition to come. It read: "Am going south. Amundsen." Initially Amundsen had been outfitting for a northern expedition across the Arctic Ocean in the Norwegian explorer Fridtjof Nansen's old ship *Fram*. Before he could set sail, however, reports told of both Frederick Cook and Robert Peary having reached the North Pole and with that news, interest in Amundsen's planned northern exploration dwindled. So did financial backing for it. Consequently he announced that rather than continue with the Arctic Ocean plan, he would round Cape Horn and sail up the coast of the Americas to Bering Strait; but instead from Madeira he cabled Scott that the two of them were in a race for the South Pole.

Actually the "race" for knowledge of the snowy extremities of earth had begun two thousand years before the time of Scott and Amundsen. It focused first on the Arctic. A Greek navigator named Pytheas in the year 325 B.C. reached as far north as Iceland and then sailed still farther until blocked by ice. The tin mines of Cornwall had been his true destination but, of an inquiring and scientific mind, he couldn't resist going farther once he heard of land to the north. He called the farthest region he reached Thule, a name that has lingered into the present.

Irishmen, Scotsmen, and Shetland Islanders may have

preceded Pytheas in the north, traveling in simple but seaworthy vessels of hide stretched over a wooden frame. Certainly after Pytheas' time Irish and Norse monks seeking escape into a life of isolated contemplation settled in the Orkney, Shetland, and Faroe islands; and some, following the route of migrating geese, eventually landed in Iceland. Sagas dated about A.D. 870 tell of their colonies and of a new wave of northern settlement by the Vikings. These newcomers had the stage of polar exploration to themselves for five centuries, until the changing climate and loss of their umbilical tie to Norway augured against them.

That same spirit of growing trade that had contributed to the neglect of the Vikings' Greenland colonies eventually quickened interest in the unknown Far North, not for itself but as a route to riches. By the 1400s educated men were accepting the Greeks' knowledge of the earth as a sphere, and the concept of circumnavigation was on its way. Land prevented eastward voyages to Cathay, so ships' captains decided to try westward routes. When Columbus made a landfall in Cuba he thought he had arrived in Japan, and he even sent emissaries with a gift for the emperor and a letter of greeting from their Royal Highnesses Isabella and Ferdinand. Five years later John Cabot, a Genoese from Bristol, also sailed westward to reach the East. He arrived in Newfoundland and Labrador, realized that their coasts couldn't possibly be outriders of Cathay, and acknowledged that a previously unknown continent must lie between Europe and the spices and silks of the Orient. The route westward was blocked by land, Cabot reported, even as the route eastward.

Perhaps the way lay to the north. By the sixteenth century successive arctic expeditions left from England to find out. Each sought to expand trade and enrich the home island. Backing came from the Royal Court, the City of London financiers, and wealthy merchants. Specialists and technicians offered advice on geography and navigation. Their only real question lay in whether the Orient might best be found via a northwest passage or one to the northeast. The northeast route held favor. A widely accepted, although fallacious, map showed it as best; furthermore, the waters were familiar to British seamen at least as far as North Cape, about five degrees beyond the Arctic Circle, and they knew that Russian fishermen regularly sailed eastward to the River Ob and possibly beyond. Also they believed that "civill people" lived along the coast, and this gave hope for brisk trade opportunities en route to Cathay, however distant it might prove to be. In comparison, the northwest route seemed unpromising. Cabot's experience was discouraging and, although Jacques Cartier returned from Canada and discovery of the St. Lawrence in 1536, his report was not published for nearly thirty years.

Whichever route was tried, the sixteenth-century voyagers outfitted with utmost care. The expedition of Sir Hugh Willoughby in 1553 furnishes an example. Backing came from "The Mysterie and Companie of the Marchants Adventurers for the Discoverie of Regions, Dominions, Island, and Places unknowen." Lead sheathed the bottoms of the vessels to repel the voracious attack of worms believed one of the main hazards of the torrid waters off Cathay—this for what in reality proved the misty ice-choked domain of polar bears. Instructions gave Willoughby authority to act upon any chance for trade, but discreetly so. "If the people [encountered along the way] shall appear gathering of stones, gold, metal or other like on the sand," the instructions read, "your pinnaces may draw nigh, marking what things they gather, using and playing upon the drum or such other instruments as may allure them to harkening, to fantasy or desire to see and hear your instruments and voices. But keep you out of danger, and show to them no sign of rigour or hostility." Also there was to be "no blaspheming of God, or detestible swearing . . . in any ship, nor communication of ribaldry, filthy tales, dicing, card tabling nor other devilish game."

In 1554, a year after they had sailed, the bodies of Willoughby and those of all the officers and men of his command ship, *Bona Esperanza*, were found near Murmansk on the Kola Peninsula, barely beyond the Finnish border. The "torrid seas" had given Willoughby a frigid reception. Blown off course, he had made landfall as best he could in a raging blizzard and resigned himself to wintering over. Death came, probably, from scurvy. With the bodies lay a note in Willoughby's hand telling of snow and hail from the first and of life along that desolate white coast only in the form of foxes, bears, and seals, with no trace of humans let alone of opportunity for trade. The route Willoughby had hoped to establish actually remained unnavigated in its entirety for nearly a century after his attempt. In 1648, during a rare ice-free summer, the Russian Cossack Semen Dezhnev seemingly reached the Bering Strait via the north Siberian coast. Some authorities question the authenticity of the evidence, however, and the final claim for the Northeast Passage may belong to the late nineteenth century when Nils A. E. Nordenskjöld sailed from Kolyna to Okhotsk in 1878–79.

Willoughby failed tragically in his attempt, but his expedition was not altogether futile. His second ship, *Edward Bonaventura*, commanded by Richard Chancellor, reached close to present-day Archangel, and there the party chanced to meet emissaries of the czar, Ivan the Terrible, who brought them by sledge across the snowy reaches to Moscow, 1,500 miles. As an outgrowth of this acquaintance a long and profitable trade between England and Russia was formed. City of London men viewed this as enormous achievement, but for geographers and cartographers it was insufficient. They wanted to fill blanks on the map, as well as in the economy. To this end the northwest route began growing in favor. Martin Frobisher was one of the first to make the search. He went seeking new geographic knowledge, yet also thirsting in the old way for trade, which remained an incentive for financial backing. On August 19, 1576, two months after leaving England, this expedition reached the southeastern shore of Baffin Island and made contact with Eskimos. By trickery Frobisher brought one aboard ship and took him back to England partly as proof of arrival in exotic northern lands. "And so they came to London with their ship," a contemporary report tells, ". . . bringing with them their strange man and his Bote [a skin kayak], which was such a wonder unto the whole city, and to the rest of the Realm that heard of it, as seemed never to have happened the like great matter to any man's knowledge." Thus did Londoners first meet the people of the snows whom they believed to live along the route to Cathay. Within days the poor Eskimo died of a cold, but Frobisher insisted that his presence gave evidence that the new land his expedition had reached must be the entry of the long-sought Northwest Passage.

The following year Frobisher commanded a second voyage north, this time returning with his holds filled with two hundred tons of supposed gold ore. Success led to a third expedition of fifteen ships which sailed from England on May 31, 1578, but this time the promise proved illusory. Two months later, off Baffin Island, "there fell so much snow with such bitter air that we could scarce see one another for the same, nor open our eyes to handle our ropes and sails." The snow soaked the men's clothes and shrouded the ships' decks and lines. This greatly "discouraged some of the poor men, who had not experienced the like before, every man persuading himself that the winter there must needs be extreme, where there be found so unseasonable a summer." The storm was a harbinger of disaster, for when Frobisher arrived home again his only welcome was word that the prized ore from his previous voyage was worthless iron pyrite. An inept, or unscrupulous, assayer had been wrong ever to identify the first black stones as gold. The news meant that Frobisher's present cargo of thirteen hundred tons of ore also was worthless, a shattering substitute for the glory he had expected and a bitter end for his voyages.

Public fascination with the north continued strong, however, despite Frobisher's folly in mining there. William Barents, a Dutchman, led three successive arctic voyages commencing in 1594, and his experiences were reported in England "as so strange and wonderful that the like hath never been heard of before." Actually great difficulty had been part of their "wonderful" venture from the outset, and on the third voyage Barents' ship suffered such extreme damage by ice that his men used the timber from the forecastle to build a hut ashore. They furnished it with a lamp fueled with oil from the fat of "cruell bears" and for comfort from the long darkness and the endless snows, they fashioned a Turkish bath out of a wine barrel. Winter treated them badly in spite of these efforts. Blizzards howled incessantly, and the cold stopped the men's clock and froze their wine and their bed sheets. Eventually they abandoned their shelter and trekked futilely from one channel in the ice to the next, dragging heavily laden boats behind them. Three hundred years later Norwegian sealers found the little hut with its cooking pots, swords, gun barrels, flutes, drumsticks, charts, books, and chiming clock intact. A sea chest still held Renaissance prints and engravings of Juno and Venus together with depictions of biblical stories. All were intended as gifts for the men of Cathay.

In a sense Barents' expeditions marked the transition from one era of polar exploration to the next. The map still was mostly blank for the high latitudes of both north and south, but knowledge at last was beginning to replace fantasy. Men from Europe had contacted the Eskimos, had wintered in the north, and knew enough about it to realize that as a route to the riches of the East it was neither short nor easy.

Through the seventeenth century and into the eighteenth, knowledge continued to grow. A breakthrough in understanding magnetism enormously boosted the effectiveness of exploration techniques, and ways of preventing and treating scurvy at last were discovered. England, France, Holland, and Russia each hoped to outdo the other in trade and in glory, and to that end mounted expeditions. One of the most significant was that of Vitus Bering, a Dane belonging to the Imperial Russian Navy, who journeyed by horse-drawn sledge from St. Petersburg to Okhotsk on the Siberian Pacific and from there coasted the Kamchatka and Anadyr peninsulas in 1729. This led ultimately to establishment of the Russian-American Company, which began operation on the American side of the Pacific by the end of the century. Supply was exorbitantly costly, however, and administration garbled because of the vast distance separating the Aleutians and Alaska mainland coast from the Russian capital.

Still, the company's charter granted a monopoly to trade and hunt, and also to explore and occupy any land unclaimed by other nations. Russian accounts of the time tell that the system of trade was for the Russians to step forward and lay out goods on the snow, then retire. Eskimos would come and look over the array of metal buttons, knives, glass beads, coral, and pearl enamel; then place beside them as many furs, or adzes, or walrus-tusk carvings as they considered appropriate. That done, they stepped back and the Russians again advanced. If the exchange was acceptable, the Europeans picked up the Eskimo offerings and left; then the Eskimos followed suit. If not acceptable, the two sides seesawed back and forth, adding a skin here or withdrawing a knife or a string of beads there. The Eskimos were reported as sharp traders, cited by one Russian as "proof that one cannot make out better by receiving than by giving."

In England, the British Admiralty of the early nineteenth century was looking for new missions and found one in northern exploration. Officers and jack tars sailed to the far regions of the earth filled with the confidence of the Victorian Age. Their uniforms stayed with established tradition wherever the navy ships went, wholly disregarding all relationship between clothing and location. Stewards set tables with cut glass and silver for the officers, heavy plate for the men. Weekly humor magazines produced by expedition personnel and the latest London stage shows, also expedition-performed, kept morale high. Naval discipline ruled at all times and, more than discipline, there also was the naval dedication with which Sir Charles Wright still spoke more than sixty years later as he told of the Scott antarctic expedition.

One of the important ventures of the time was Edward Parry's. He departed from the Thames on May 11, 1819, heading for Kamchatka via the northwest but destined instead to hear the crack of his ship's timbers as winter ice held him fast off the east coast of Canada with "not an object to be seen on which the eye could rest with pleasure." In 1821, Parry headed northwest on a second voyage, and in 1824 on a third. He added overland exploration to coastal charting, specially equipping his ships' boats with iron runners and carrying lightweight sledges and provisions for two and one-half months of oversnow travel. Snowblindness forced the party to move only at night, and pack ice kept drifting away with them at a rate faster than the progress they could make northward. Yet by July 26, 1827, Parry reached to within 435 miles of the Pole, a northern record that stood for 50 years.

More new equipment and methods were tested—worthwhile ones and foolish ones, half workable ones and half impossible ones. Men built their ships with heavy oak ribs and planks to resist the ice, so much reinforcing that one old salt is quoted as complaining to his captain: "Lord, Sir! You would think by the quantity of wood they are putting into the ships that the dock-yard mates believed they could stop the Almighty from moving the floes in Baffin's Bay!" Methods of propulsion became innovative. One expedition outfitted a paddle steamer off the Calais-to-Dover run for the north. Its engine took up most of the hold, its boilers leaked, and its top speed was three miles per hour until the captain gave up the slow churning and switched to sail.

More successfully, the Arctic became a testing ground for the new screw propeller, a great step forward in marine design although not an immediate triumph. Sir John Franklin's expedition, an ironically important one, helped to pioneer use of the propeller. His ships even were equipped with the added refinement of being able to raise their propellers to protect them from the ice. Converted railroad engines provided power, leading to a comment from one of Franklin's men that "Our engine . . . has a funnel the same size and height as it had on the railroad and makes the same dreadful puffings and screamings, and will astonish the Esquimaux not a little." In late spring 1845, the expedition cleared the shores of Britain and by the end of June crossed the Arctic Circle off Greenland's west coast, intending to pass into Bering Strait by autumn. Franklin, his men, and the public expected a great deal of new knowledge to be gained. But it was destined to happen only indirectly as would-be rescuers searched vainly for survivors. In July the captain of a whaling ship that was moored to an iceberg spoke with the men of the Franklin expedition, then watched the two vessels, *Erebus* and *Terror*, steam off. No one outside the Arctic ever saw them again. The entire expedition—134 men—vanished and little trace ever was found despite more than two decades of looking.

In all about forty searches were made. Each party charted and mapped the maze of arctic islands and waterways as they went and in this way contributed to basic understanding of the Arctic. The primary motivation of all, however, was hope of finding Franklin or at least of learning his fate. Everything anyone could think of was tried. One expedition filled balloons with hydrogen and released them over the snowy wastes with dangling pieces of brightly colored paper attached. The idea was that these would give Franklin's men the position of the relief party "should [the balloons] happily fall near the poor fellows." Another expedition sent carrier pigeons flying over the endless white expanses with notes from wives and sweethearts, but none brought back a message from the stranded party. Arctic foxes were tried too. They were live-trapped and fitted with collars that held messages; but when the animals kept getting retrapped, this effort, as the others, was sadly given up.

In time discovery of skeletons and relics and papers let Franklin's story be pieced together. The two ships had become hopelessly caught in the ice, and their men eventually set out for the Back River, which they thought would take them into Hudson's Bay. Dreadfully wearied, malnourished, and doubtless afflicted by scurvy as well, the party broke discipline and splintered into separate groups. They were weakened horribly, yet they approached travel across the sterile snow desert of the north with preexisting, European thinking. When found, their bodies lay clothed as British seamen, neckerchiefs properly knotted and clothes brushes at hand. Several lay face down, evidently having pressed on until they fell over dead. Two skeletons lay pathetically beside a boat of seven hundred pounds mounted on an oak sled of over six hundred pounds—an impossibly punishing load to drag over the snow. Stowed aboard were the men's toothbrushes, combs, soap, towels, and sponges as well as more basic survival materials such as bullets, nails, twine, and a sailmaker's palm. Apparently one party of survivors wandered about the northwest corner of the Melville Peninsula until as late as 1862 or 1863. Eskimos later reported having seen these men but mistaken them for Barrenlands Indians, who were enemies. Consequently they offered no aid, and with the deaths of these last survivors the final curtain of tragedy rang down on the Franklin expedition.

Emphasis now shifted to the race for the Pole, as such. The search for Franklin led to recognition that the Northwest Passage was an impractical link between Europe and Asia. The geographic pole, however, that elusive northern point—at which all lines of longitude converge and the only direction is south—remained enticing and unknown, although the techniques of arctic exploration were showing steady improvement. Charles Hall, an eccentric Bostonian who sought futilely for evidence of Franklin during the 1860s, had called attention for the first time to the advantages of dressing, eating, and traveling the Eskimo way. Rather than impose oneself on the snow and ice, Hall pointed out, wisdom lay in adapting to the realities of the Arctic and seeking to fit in harmoniously. Trying to bludgeon the North into submission by sheer determination and technology was foolish. Fridtjof Nansen with three other Norwegians and two Lapps furthered these recommendations by getting Eskimo help in designing equipment, then testing it in 1882 by skiing across the Greenland ice sheet, a four-hundred-mile trip that included climbing nine-thousand-foot mountain passes.

Late in the century attempts on the Pole were made under the flags of England, America, Australia, and Norway. Many said such a prize scarcely could be worth the cost of reaching it, but no long-held goal is that simply—or rationally —dismissed. The expeditions continued. As the nineteenth century closed, the American explorer Robert E. Peary determined to stand at the Pole itself, and on a third try he succeeded. In July 1908, he steamed out of the New York harbor and ultimately shoved off across the ice on a course set west of north to compensate for drift. On April 2, 1909, he commenced the last stage of his journey. Accompanying him were Matthew Henson, a Black, and four Eskimos. The party raced the full moon and the tides, which threatened to open impossible leads in the ice—and they won. "The Pole at last," Peary wrote in his diary on April 6, 1909. "My dream and goal for twenty years." Like a conqueror from an earlier age, he claimed the entire region in the name of the President of the United States and then planted five flags into the ice: the national flag and those of the Navy League, the Red Cross, a special ensign of world "Liberty and Peace," and the Kappa Epsilon fraternity flag of Bowdoin College.

On April 7 Peary and his men started back from the Pole, and soon he cabled his triumph to the world. "This work is finished, the cap and climax of nearly four hundred years of effort, loss of life, and expenditure of fortunes by the civilized nations of the world. I am content." His contentment was short-lived, however. On arriving home he immediately found himself embroiled in a controversy of such magnitude that to this day it is not resolved fully. Dr. Frederick Cook, a polar pioneer who had been with Peary in Greenland in 1898–1902, claimed to have reached the North Pole the year before, on April 21, 1908. By the time Peary returned, feeling exultant, he found Cook's report published with the title "My Attainment of the Pole." Congress investigated. So did a special committee of the National Geographic Society. Peary emerged the victor, but the edge was forever gone from the final glory that so long had lured men north. "Polar exploration is at once the cleanest and most isolated way of having a bad time which has been devised," a colleague of Peary's wrote. Few who had experienced it would disagree.

In the North, however, the search for the Pole at least had been backed by a long acquaintance between man and Arctic. For no matter how elusive the Pole itself, or how

slow Europeans and Americans had been in adapting themselves to its white wilderness, man has lived in the Arctic
for at least thirty thousand or forty thousand years and
possibly for untold millennia more. In the Antarctic, no
such familiarity exists. Instead man and land only recently
have met. Amundsen's and Scott's achievements came as the
swift climax of only a few centuries of speculation concerning the South Pole rather than after millennia. The ab-
origines closest to Antarctica lived at Tierra del Fuego, the
large island off the tip of South America named for the fires
of the Indians, but so far as is known these people never
ventured to the continent south of them. The only legend
of such a voyage comes from a very different quarter. Poly-
nesians tell of a warrior who sailed so far that he entered
a dismal realm of fog and boundless ice. On return to the
sunny palm islands of home he recommended against anyone's traveling so far again.

Until the eighteenth century, geographers had little
knowledge of the far south latitudes, although they believed
that a huge continent spread across "the bottom of the
world." They expected it to be fertile rather than desolately
white. Great navigators from Columbus' time on added
fragments of understanding, but it wasn't until 1768 when
the great British mariner James Cook strode onto the stage
of history that scattered bits of observation at last could be
fitted together. Cook had immense drive and ability, and
at the time he began his explorations he was in his late
thirties and therefore with enough years of life remaining
to settle major questions of world geography. On his first
voyage, 1768–71, he navigated from Cape Horn to Tahiti,
circled New Zealand, coasted Australia to the east, and
returned to England by way of New Guinea. He reported
that no continent lay in the region of the fortieth parallel
and that if such a discovery ever were to be made the
search would certainly have to be made farther south. The
public, however, paid little attention to these words, preferring instead to believe that Cook had found the fabled land
of the south and that it was inhabited by people "hospitable,
ingenious, and civill." The government actually was ques-

tioned concerning its involvement with colonies in America when so clearly a more spacious and richly endowed land lay to the south. In time, Cook's full account and true viewpoint gained general acceptance, but by then he was off on his second voyage.

This time his ship was *Resolution,* 462 tons, accompanied by *Adventure,* 336 tons, and his sole purpose was exploration. On January 17, 1773, the two little vessels crossed the Antarctic Circle, the first known with certainty to have done so. Cook told thus of the day: ". . . by 4 P.M., as we were steering to the south, we observed the whole sea in a manner covered with ice . . . and as we continued to advance . . . it increased in such manner that . . . we could proceed no farther, the ice being entirely closed to the south . . . without the least appearance of any opening."

In the next few days Cook crossed the Antarctic Circle twice more, although without sighting land as he would have done at that latitude had he been in almost any other sector. Turned back by pack ice on January 26, he wrote that while perhaps it was possible to get farther south, the effort seemed rash. "It was, indeed, my opinion, as well as the opinion of most on board, that this ice extended quite to the Pole. . . . And yet I think there must be some [land] to the south beyond this ice; but if there is it can afford no better retreat for birds, or any other animals, than the ice itself, with which it must be wholly covered. I, who had ambition not only to go farther than anyone before had been, but as far as it was possible for man to go, was not sorry at meeting with this interruption [as it] shortened the dangers and hardships inseparable from the navigation of the Southern Pole regions." Unable to proceed "one inche further south," Cook turned for home.

In spite of Cook's observations most men refused to replace their long-held vision of a green southern continent with the actuality of white desolation. In part this was because Cook had mentioned seals and whales in abundance, a clear invitation to wealth as commercial seafarers saw it. Furthermore, so men reasoned, where there was ocean life there must also be gentle lands fit for human settlement.

British, French, Russian, and American governments and private companies responded to what they saw as new opportunity and, although no rush to southern waters immediately ensued, discoveries by navy men, whalers, sealers, and would-be colonizers extended an understanding of extreme southern geography.

William Smith, an Englishman who had learned the ways of whales off Greenland, was one of the commercial seafarers who helped to fill in the map. He sailed out of Valparaiso in December 1819 bound for an antarctic island and taking along livestock. He had visited the island before, yet even so clung to the general expectation of a fertile southland. His plan to graze cattle and sheep of course quickly failed, but his attempt contributed to knowledge. He sighted the antarctic mainland on February 4, 1820, and claimed it for Britain, naming it Graham Land. That same year an American sealer, Nathaniel Palmer, also sighted the same shore with the result that American maps referred to the Palmer Peninsula rather than to Graham Land. By chance, Russian claims also date from 1820—and in fact can be said to predate those of England or the United States by a few days. Baron Fabian Bellingshausen, German-born but sailing for the czar, came within sight of Graham Land/Palmer Peninsula a little before the English and American sightings, but he arrived during a snowstorm and couldn't see through the bleak whiteness to recognize his position.

In 1838 the United States Congress echoed the convictions of the Russian and British governments that a navy's concern properly should include exploration, and they authorized an ill-planned and ill-equipped antarctic expedition which Lieutenant Charles Wilkes led with remarkable effectiveness. In 1840 Wilkes brought his ships and pitifully poorly clothed and scurvy-ridden men into antarctic waters and within sight of land, thus greatly furthering American interests. That same year a French expedition commanded by Dumont d'Urville also sounded and charted south of latitude 66°33', successfully discovering land which d'Urville named Terre Adélie for "the devoted companion who has three times consented to a painful separation in order to

allow me to accomplish my plans for distant exploration."
Pushing through a penguin colony, the men unfurled the
Tricolor and "by wholly pacific conquest" considered them-
selves on French soil.

A third expedition at about the same time was commanded by James Clark Ross, the Scotsman who a few
years earlier had discovered the north magnetic pole and
to whom the British Admiralty now gave orders to sail in
search of the south magnetic pole. To Ross's immense joy as
a patriot he sighted distant land more southerly than that
reported by mariners sailing under other flags, and this fact,
he proudly pointed out, "restored to England the honour of
discovery of southernmost land." Ross's party saw and named
Mount Erebus and Mount Terror, volcanoes that astonished
them in the seemingly endless snow and ice. Erebus, 12,400
feet, was in eruption, "a most grand spectacle." The party
also found Ross Island, which they called High Island, and
the Ross Ice Shelf, which they called Victoria Barrier, adding the comment that a man "might with equal chance of
success try to sail through the Cliffs of Dover as to penetrate
such a mass."

In a sense Ross, who completed his voyage immediately
after those of Wilkes and d'Urville, closed the era of major
antarctic exploration by sea and raised the curtain on the
land exploration that would culminate with Scott and
Amundsen and lead to our own day of sophisticated research.
In another sense Ross's reports added specific impetus to
the probes already begun by seal and whale hunters. He
commented on the abundance of life encountered in antarctic waters and thereby quickened the thirst for riches to be
made from sealskin and whalebone and from rendering the
oil of the animals' carcasses. With the opening of major
whaling and sealing came the beginning of detailed geographic knowledge as replacement for the speculation of
previous centuries. The white realm of the south, as that of
the north, was becoming known. By 1912 both poles had
been won; man had reached the ends of the earth at last.

CHAPTER

4

ARCTIC AND ANTARCTIC

THE TWO POLES DIFFER GREATLY. THE ANCIENTS WHO THOUGHT of the Far North as the land of Arktos, the Great Bear (Big Dipper) were right to name its counterpart the Ant-arctic, for in many ways the southern polar region opposes the Arctic. It is an ice-covered continent set in a partly frozen ocean, whereas the Arctic is an ocean choked with drifting ice, dotted with islands, and nearly ringed by continents. The North Pole must be located by instruments because its character changes from day to day as the pack ice clashes and grinds in its slow clockwise drift. Because of water's moderating effect the mean annual temperature at the Pole is a mere −9° F., and the winter mean for the Arctic as a whole varies from −20° to −40° depending on elevation and latitude. Conditions are such that a wide range of life can thrive, including large land mammals.

Not so the Antarctic where the annual mean temperature of the Pole is −58° F., and in the entire region the biggest land animal is a wingless fly less than a quarter-inch long. Antarctica is a continent capped by so much glacial ice that the sheer weight in places is pushing bedrock to sea level and below. Nowhere else on earth is there so much ice. It overrides the land as an incredible blanket up to three miles thick. Only the greatest peaks rise free. The elevation of the continent as a whole averages about six thousand feet, which is higher than any other continent, and peaks soar to nineteen thousand feet.

This height of the Antarctic almost is matched in magnitude by the depression of the arctic basin, which averages 4,200 feet and has a maximum depth of 17,500 feet. This is much more than men at first supposed. Late in the nineteenth century Nansen sailed north to test ideas of arctic drift by letting his sturdy little vessel *Fram* freeze into the pack ice and be carried with it by the slow circling of the current. At the time nobody knew most of the Arctic firsthand, and geographers disagreed as to its character. With *Fram*, Nansen firmly established the wholly oceanous nature of the highest north latitudes, which were expected but not proven, and also their great depths, which weren't anticipated. To make soundings he had taken only hand winches aboard and such short reels of bronze wire and steel cable that, to his astonishment, individual strands had to be separated, retwisted, and the pieces then soldered together in order to reach bottom. This newly fashioned line was lightweight, but it served. Eleven successful soundings of from 11,000 to 12,300 feet were made as the vessel drifted with the ice. Finding this depth added more to knowledge of the north polar basin than any of the other observations made during the entire expedition. *Fram* reached 86°14′ north latitude, at the time farther north than any other ship had gone. The vessel today is on display in Oslo, along with the drawings for its construction, which date from 1892. *Fram* was built specifically for the ice. You can go aboard and in the hold see the huge oak timbers and the cross bracing, successfully designed to withstand the crushing force of the ice.

The tragic voyage of the *Jeannette* a few years before Nansen sailed had established the clockwise drift of the ice. The *New York Herald* sponsored the expedition, which was commanded by Lieutenant George Washington De Long of the United States Navy. The same paper earlier had built circulation by sending Stanley to look for Livingston in Africa and also by sponsoring a search for relics of Franklin's ill-fated party. De Long left San Francisco on July 8, 1879, and by September 6 his vessel was frozen helplessly in the ice. For twenty months it drifted with the pack; then the ice opened, only to close back and crush the already battered ship. De Long and most of his men perished, but not even death stopped the accomplishment of their expedition. Three years after the disaster, wreckage unmistakably from *Jeannette* was found on an ice floe off the southwest coast of Greenland, the opposite side of the polar basin from where ice had smashed the ship. Several objects were present, including clothing marked with the names of crewmen. Discovery of these items, along with the fact that driftwood believed to have come from Siberia also had been found in Greenland, made it clear that a great current rotated around the polar sea.

Nansen knew of this and decided to test the drift empirically and directly, hoping it would take him to the Pole itself. That didn't happen, but his technique of making observations while drifting was to become standard. In 1918 the American anthropologist Vilhjalmur Stefansson manned the first scientific station to be established on an ice floe, drifting with it for eight months to record data on the pack and the tides. Today such ice-island stations are regularly maintained by both the United States and the Soviet Union, and have been for forty years.

The 1920s and 1930s brought real understanding of the nature of both polar regions as men increasingly penetrated them. As early as 1926 Amundsen flew over the North Pole in a dirigible, thus becoming the first man ever to attain both poles. Two years after that flight Sir Hubert Wilkins made the first airplane crossing of the Arctic Ocean. In another three years, in 1931, he made a submarine attempt for the Pole but was forced to turn back because of damage to the diving gear. In Antarctica similar advances were made. Admiral Richard E. Byrd flew his Ford trimotor monoplane from Little America to the Pole and back in one day—sixteen hundred miles—in 1929. To reduce its weight, celluloid was substituted for glass in the windows, saving 80 pounds; the cabin was untrimmed, saving 155 pounds; and the plane's skin was a special lightweight one that saved another 235 pounds. Despite the cumbersome sound, this was a great improvement in equipment. Even that early, polar travel was easing enough to let men turn to searching for knowledge without being encumbered by the problems of survival, which had characterized earlier exploration. By 1962, on return from a stay in Antarctica to study the upper atmosphere, Sir Charles Wright remarked, "The progress that has been made in transport and in instruments is staggering. It is quite an experience to fly up the Beardmore Glacier toward the Pole in a couple of hours, while remembering the two months of hard sledging that the journey once required."

Yet even with the overall character of the white nether-reaches of earth now known, certain basic statements remain complex. For instance, the size of the polar regions can be given generally as about five and one-half million square miles for the Arctic Ocean, closely matching the slightly more than five-million-square-mile expanse of the antarctic continent. But try to define the boundaries of the two and a problem arises. The Arctic Circle at 66°33' north latitude is of course the exact counterpart of the Antarctic Circle at 66°33' south latitude, but these imaginary lines really aren't satisfactory delimitations. Each girdles the earth at the position of the midnight sun, marking the latitude where for one twenty-four-hour day each winter the sun never comes above the horizon and, conversely, where it never sets for one day each summer. This means that the Arctic and Antarctic circles are relevant to astronomical events but say nothing about the topography of the earth or the nature of its habitats.

Climate considerations are more valid. The widely recognized Köppen system of climate defines Arctic and Antarctic as regions where the warmest month averages below 50° F., and the coldest month never rises above freezing. From a biological standpoint this line, in the Arctic, corresponds fairly closely with the northern limit of trees—although defining a tree isn't easy at that latitude. A genus that clearly produces trees elsewhere may in the Arctic amount to nothing but brush or even less. Certain willows stand no more than an inch high when fully mature. The Antarctic has no trees by any definition. Lichens and mosses are its major vegetation, and they are sparse.

Another approach to defining the polar regions is to use the limit of floating ice as the beginning of the Arctic or Antarctic. The concept is reasonably appropriate in the polar south but not in the north since it leaves out parts of the Greenland coast and the Barents Sea along with all of interior Siberia, Alaska, northern Canada, and Greenland— yet the coldest temperatures within the Arctic are recorded not at the North Pole but in the highlands of eastern Siberia where the winter minimum has sunk as low as −93° F. In the Antarctic ice does form a distinct boundary. By itself, the pack amounts to a virtual physical barrier, and in addition its outer edge coincides with the point at which warm salty water flowing down from the north merges with cold Antarctic water, chilled and freshened by ice melt. Where the two meet, a great upwelling takes place as the cold water sinks and the warm water rises to replace it. This stirs nitrates and phosphates to the surface, enriching the nutrient broth of the water for the few species of plankton adapted to the dominating cold. Unadapted species from the north die, their crucial temperature and salinity requirements no longer met. The presence of the plankton, whether living or dead, makes the convergence zone a rich feeding ground for birds, fish, seals, and whales as well as for deep-water sponges, mollusks, and other filter feeders nourished by the organic rain from above. So distinct are the effects of the waters' meeting that a mariner usually knows without looking at his chart that he has arrived at about 50° south latitude in the Atlantic or Indian Ocean, and between 55° and 62° in the Pacific. A gray bank of fog will lie ahead, and crossing into it voyagers soon find the fog turning to mist and the mist to snow. Such wet weather is all but perpetual as vapor held by warm moist air increasingly surrenders to the bitterly cold southern air.

In both polar regions sea ice forms when surface water temperatures drop to about 28° F. Snow weights the thin new ice and causes it to sag, which sends water flooding over it. This quickly freezes, thickening the ice. In the north, pack ice amounts to a shifting, grinding lid that covers the ocean several feet thick from fall into spring. In the Antarctic, pack ice is very different. There the winter pack rims the continent for nearly five hundred miles and is a combination of sea ice plus immense icebergs calved from glaciers. The summer pack retreats to a comparatively narrow band except for semipermanent ice in much of the Ross and Weddell seas and off the Pacific side of Lesser Antarctica. Comparing the seasonal change, Admiral Byrd once wrote that although the antarctic pack ice often is a dangerous enemy in midsummer, its gauntlet generally can be run. But in winter, "it reigns supreme. Probably all the navies of the world together could not batter their way through."

In addition to the pack, the Antarctic has two other types of ice—and a total ice volume that exceeds the Arctic's by about eight times. Aside from pack ice there also are shelf ice and glaciers. The shelf ice forms as layers of snow blow from the land onto the frozen sea and compress. This ice stays attached to the land because it is continually replenished from that direction. The best known example is the gigantic Ross Ice Shelf which covers inner McMurdo Sound, an area the size of France. This ice stretches nearly 400 miles across the sound, and at its outer edge is 350 miles from the innermost shore. Its glistening cliffs tower 200 feet above the water and reach four times that far beneath the surface. In places glaciers augment shelf ice. They stretch across the land and into the water like gigantic frozen tongues floating at their lower ends. The Beardmore and Axel Heiberg glaciers contribute in this way to the Ross Ice Shelf, and antarctic glaciers as a whole account for the greatest part of the ice that whitens the entire south polar region.

High in elevation and far removed from a moderating effect of the ocean, the continent experiences temperatures as low as −127° F., the coldest known anywhere in the world. This reading was taken in 1960 at Vostok, a Soviet research station situated on the ice cap eleven thousand feet above sea level. A standing rule there is that men should be outdoors for only fifteen minutes at a time when the temperature falls to −112° F.; when it drops to −121° F. or lower, the time is cut to ten minutes. The buddy system of course prevails, as it does at all antarctic stations: No man ventures forth alone. When going out, multiple layers of clothing, including furs, are standard attire, and in addition men carry battery heaters to warm hands, feet, and chests. They also wear face masks which are fitted with hoses to allow breathing warm air from inside clothing. Thus equipped, they carry out their observations.

Actually, despite harsh conditions, Antarctica in many ways serves as a more convenient high-latitude observatory than the Arctic Ocean is ever likely to be. As Sir Charles put it, the southern ice cap remains wholly inhospitable, but it at least offers a solid base for permanent stations, and its most remote reaches are easier to get to than corresponding points in the Far North. This has made the Antarctic the primary location for study of many characteristics common to both polar regions. In 1961 nineteen nations signed a treaty of cooperative international scientific study in the Antarctic with all individual political, economic, and military gain forever forsworn. Among the ensuing studies have been several concerned with peculiarities of geomagnetism and the upper atmosphere, crucially different at the poles than in the middle latitudes. Lines of magnetic force, which arc out from the poles, set the Arctic and Antarctic apart from the rest of the earth in several ways. Most spectacular of these are auroras—the moving curtains and bands of eerie light that glow in the night sky of the high latitudes (and are detectable also in the daytime by using radar). The best explanation of auroras is that they occur as protons and electrons from the sun bombard oxygen and nitrogen molecules in the upper atmosphere. Because of the geomagnetic field, incoming electrically charged particles are spiraled toward the poles, and there the phenomenon appears in its fullest glory. So intense is the effect that all peoples at high latitudes know the sight. Nansen, wintering off the New Siberian Islands, wrote: "In the midst of this silent, silvery ice-world, the northern lights flash in matchless power and beauty over the sky in all colors of the rainbow." The light can be truly brilliant. An exceptional aurora even has been reported from Mexico City, far beyond the usual range of visibility.

The Van Allen radiation belt also is a magnetically controlled dance floor for high-energy solar particles but, the converse of auroras, its effects are minimal at the poles. Consequently the Antarctic has been suggested as a possible location for launching manned flights into outer space. Outward bound or touching back to earth, astronauts would receive minimum exposure to the dangers of Van Allen radiation if they exited and entered the atmosphere at the Antarctic. The Arctic would have equal advantages but is impractical since the region of the Pole is ocean.

Such potential utilization carries the clear hallmark of the present and future, but other applications of peculiarities in the polar atmosphere are ancient. Arctic Eskimos long have read a "sky map" to assess terrain beyond range of normal vision. Such maps are simply reflections of ice, water, and land surfaces registered on the clouds, but from them knowledgeable men can tell whether a lead of open water is within range of the village or, when hunting, can study the points of land and the bays ahead and decide the best place to camp. Leads reflect as dark streaks, winter sea ice as dull white expanses, and snow-covered land as brilliant white expanses. Possibly the Celts from the Faroe Islands, who journeyed round trip to Iceland in A.D. 800, had sky-map information regarding land to the far northwest. The Vatna Glacier, which rises 6,200 feet above the southeast shore of Iceland, could cast its reflection onto clouds and be interpreted correctly as indication of land beyond the horizon. The necessary observational skill is likely for a people accustomed to the northern sky. The meteorological phenomenon is fully plausible, although its effect would be apparent only in the latter part of the 230-mile voyage from the Faroes to Iceland.

Better than a simple reflection for conveying information would be a mirage—and arctic mirages are not unusual. They occur when cool, and therefore dense, air is overlain by warmer air. The resulting inversion causes light to refract in an arc and, in effect, give an observer a look beyond the horizon. An object such as the gleaming white of a high snow slope will appear vertically displaced from its true location and may be seen as a mirage from a much greater distance than is true of a reflection. For the questing Celts— or for Eric the Red responding to persistent tales of something beyond the horizon—it may have been mirages that provided advance information. Given a temperature inversion with a gradient of less than one degree Fahrenheit, an object no larger than a Viking longboat could be "seen" in a mirage from a distance more than seven times that without the inversion. Instead of losing sight of such a vessel over the horizon at a little more than four miles, a man standing at the height of a similar vessel can see it for about thirty miles. If he is higher than that, the distance he can see will increase proportionately.

Mirages, as auroras, are common to both poles. So are other peculiarities of light such as full halos around the sun and bright spots at opposite sides of such halos, called sun dogs. These both are caused by the bending and mirroring of the sun's rays as they pass through clouds of tiny ice crystals. Another condition common at the poles is a whiteout when light reflects back and forth between snow-covered land and an overcast sky, causing the horizon to vanish and leaving skiers and pilots disoriented in an unbroken sea of white with no landmarks for guides.

Both polar regions are dominated by ice and snow and cold—yet both are deserts. Precipitation is scant. Less falls in the Arctic and Antarctic than in Arizona's Sonoran Desert; the yearly polar average resembles that for parts of California's Mojave Desert. This seeming incongruity has a simple explanation: The moisture capacity of air increases with warmth and, at the poles, there is very little warmth. Air at −40° F. holds one-tenth what it can at 50° above zero, and this alone augurs against much polar precipitation. It means that the Arctic and Antarctic are snowy not because of what falls each winter but because of what has fallen already and not melted. In the central Arctic and on the Greenland ice cap yearly precipitation totals less than five inches. Over the whole of the antarctic continent the amount is five or six inches of water equivalent (from a snowfall of about sixteen inches).

This lack of available moisture greatly restricts life. The few truly polar land mammals that exist, such as polar bears and arctic foxes, belong exclusively to the north. They guard against the cold with thick fur and various adaptations in behavior. They get moisture from their food, much of which comes directly from the sea. Antarctica has no terrestrial animal life except for invertebrates such as the fly, plus springtails, mites, and similar simple life-forms. The seals and penguins of the Antarctic really belong to the water. Plant life also is scarce and stunted. Only three species of flowering plants live in all the vastness of the continent, compared with nearly one thousand for the Arctic. There are two kinds of wire grass and an herb related to carnations. More common are algae, lichens, and mosses which crust rocks and tuft sterile ground. Their nutrients come from saltwater spray and from the droppings of birds that feed in the sea but come ashore to nest. In almost all of Antarctica the only source of moisture for plants is melting snow, and the incessant, utterly dry winds rapidly evaporate what little of this there is. Light summer snowfalls melt at about the thirty-five-hundred-foot level and leave a fleeting legacy of moisture that becomes clearly marked by a line of crustaceous lichens. The drainage patterns of longer-lasting snowbanks are faintly outlined by rosettes of green algae. A few mosses have managed to spread around the coastal rim of the continent, where moisture is relatively abundant, and liverworts also are present on the Antarctic Peninsula, which is characterized by summer mildness and frequent drizzle.

Among the best growth situations are minute spaces between rocks and the ice or snow that abuts them. Such niches provide both warmth and moisture. Three hours of summer sunshine can raise the air temperature within them from zero into the low eighties F. and melt enough snow to create fragile zones of moisture. Additional windblown snow sifting in contributes still more life-giving liquid. Springtails and mites, the most abundant and widespread of antarctic land fauna, also utilize the minuscule warmth of rocks. They gather beneath them and under clumps of moss and lichens, alternating between activity and dormancy, and knowing no fixed cycle of seasonal growth or breeding periods but only the fickle coming and going of the warmth.

Around the rim of the continent but not in the interior, meltwater pools harbor populations of microscopic rotifers, diatoms, and water bears, which are primitive relatives of scorpions, spiders, mites, and ticks. On the Antarctic Peninsula tiny crustaceans called fairy shrimp are present, as well, and brackish ponds there are the breeding grounds of the wingless flies.

Antarctic soil is scarce and poor. Much of what there is holds no organic carbon except near penguin rookeries where guano provides enrichment, or where decomposition from patches of lichen or moss has left a bit of humus. Bacteria are present within the soil, however, including two strains capable of fixing nitrogen from the air. Some bacteria probably are transported on dust that blows to Antarctica from other continents, but they are present in sufficient numbers to indicate a true population that belongs specifically there. Even permanently frozen sediments fourteen hundred feet below the surface have been found to hold bacteria, the oldest known anywhere. Their depth within the earth indicates at least ten thousand years of existence and possibly as much as one million years. Remarkably, when brought to a laboratory and placed in a nutrient solution some have become active.

The largest stretches of ice-free soil on the continent are in the dry valleys near McMurdo Sound (one of them named for Sir Charles Wright). These valleys lie in the lee of a mountain range, evidently carved by glaciers, although the ice is long gone. Jumbles of boulders alternate with sand and square miles of polygons that resemble the well-known patterned ground of the Arctic (and are caused by the same freeze-thaw action of waterlogged silt and gravel). A few streams flow through the dry valleys, and there are lakes, many of them ringed by successive shoreline terraces left as water levels have changed. Each lake has its own unique character. One is covered by ice ten feet thick; yet, for reasons unknown, the bottom stays about 80° F. Volcanic heat once was believed to be the cause, but underlying strata have proven cold. Another lake doesn't freeze despite extremely low winter temperatures. It is too salty.

Most surprising, the remains of seals have been found in some of the valleys. Men on Scott's first expedition came across two such Weddell seals more than fifty miles inland and at an elevation of five thousand feet. The second Scott expedition found thirteen mummified seals on the Ferrar Glacier and its moraines, together with the fleshless bones from so many others as to suggest their presence over a long period. More recently, American researchers found ninety crab-eater seal mummies in Taylor Dry Valley. These are especially mystifying. Weddell seals occasionally are seen as far as thirty-five miles from the coast, usually aged bulls that seemingly go inland to escape sea leopards and killer whales. But crab-eaters are practically unknown away from the shore except for this discovery. All the mummified seals are believed to have traveled inland centuries ago when ice floored the valleys. Radiocarbon dating shows them to be as much as twenty-five hundred years old, although the dates are somewhat suspect because upwelling of ancient carbon could have entered the food chain of the ocean and made its way into the seals' tissues. Probably it never can be known just when or why the seals made the journeys, but their mummies and bones in the silent, cold, dry valleys emphasize the essential sterility of the continent. Found inland they seem drastically out of place. In Antarctica only the ocean can support vertebrate life, not the land.

Among the aquatic life-forms of the Antarctic are penguins, birds peculiar to the southern polar realm which come ashore only to nest. Forty-three bird species are known south of the Antarctic Convergence, seven of them penguins. The others include twenty-four species of petrel, five species of gulls, skuas, and terns, two cormorants, two ducks, two wading birds, and one songbird, a pipit. These are all water birds and all but the last five are seabirds.

The Antarctic Ocean exceeds the yield from the world's most fertile agricultural land in carbohydrates, fats, and proteins. Its food production is quadruple that of any of the world's other oceans, so rich that men are considering turning from the hunting of whales, now overkilled, to harvesting the chief food that has sustained them. This is krill, small crustaceans that feed on diatoms. About two inches long, these clawless "lobsters" reach a density as high as one for each cubic inch of water. Fecundity of this magnitude readily supports the shadowy forms of fishes and mammals swimming and feeding within the water and of birds by the hundreds of thousands wheeling and calling above the waves, then plunging within to feed. But the teeming life belongs only to the antarctic summer, to the period of sunlight when the microscopic plants of the phytoplankton can photosynthesize and thereby form the key link in the food chain. When the sun disappears below the horizon most of the conspicuous life-forms of the Antarctic also vanish. The minute sink into dormancy, and most of the large forms migrate. A very few tough it out.

Even those species that forsake the antarctic winter find the southern continent far from temperate. The mildness of its summer season is unendurably harsh by most standards. For example, silver-gray petrels nesting on exposed rock faces often work with their beaks for hours to clear snow from the crevices where they have laid their eggs and, when blizzards delay the return of adults feeding at sea, hatchlings starve or freeze to death. Elephant seals breeding among the grass tussocks of South Georgia Island, which lies north of the true Antarctic, also are plagued. The warmth of their bodies sometimes melts holes in the snow crusting between tussocks, and pups drop for several feet to lie helplessly trapped and beyond any chance of suckling. Such is the legacy of the mild season. In winter the elephant seals and the petrels are gone.

Weddell seals, on the other hand, stay. They maintain breathing holes through the ice with their teeth, although not by biting. Herbert Ponting, cinematographer on Scott's second expedition, filmed the process. The seals open their jaws and use their teeth like ice picks, driving with their shoulders and swinging their heads to chisel holes open. In winter when the thickness of the ice reaches ten feet and more, they use the same method to haul out. They can't shoot up in their holes high enough to get their flippers over the edge when the ice is that thick, so they rasp an inclined trough with their teeth and work their way up. Once on top they dry off by rolling in the snow. Otherwise they soon would be sheathed in ice. For the most part Weddell seals meet the antarctic winter by avoiding its fury on land. They stay under the ice, or among the hummocks and caverns of its pressure ridges. In this way they escape the wind and the temperature fluctuations of the terrestrial world.

This isn't true of emperor penguins. For them winter is nesting time—the coldest, darkest, stormiest part of the year. Their chicks need months to fledge fully, and if they hatched in summer their plumage still would not be complete when autumn cold set in. They might die. But by hatching in winter they have the benefit of parental brooding through the ferociously hostile period of the year. By the time they have grown too big to fit under a parent's body the warm season has come, and they can be left safely while the adults

Snow changes to glacier ice in places where more snow falls in winter
than melts in summer. Often thousands of feet deep, such ice flows
downslope like a frozen river. Crevasses split the surface where basal ice
drops over cliffs. Some are narrow enough to leap across, or are
partially bridged by snow. Others gape sixty to seventy feet wide
or more. Most taper to mere cracks at depths of about 100 feet.
(Ruth Kirk)

The Antarctic's Byrd Glacier stretches for about ninety miles,
draining from the continent's central icesheet to the Ross Iceshelf.
Glaciers lock three quarters of the Earth's fresh water in their grip.
Those in Antarctica blanket an area equivalent to Canada and the
United States combined. *(U.S. Navy for the U.S. Geological Survey)*

Roald Amundsen led a 1911 Norwegian expedition to Antarctica, the first to reach the South Pole *(below)*. The men traveled 400 miles from base camp, first riding dogsleds, then pulled on skis by the dogs. *(Norsk Polarinstitutt)*

Sir Charles Wright, glaciologist on the 1911 British South Pole
expedition, returned to one of the party's abandoned huts fifty years
later with Les Quartermain. En route to the pole, Scott's men relied
at first on horses, then had to man-haul their sledges.
(National Science Foundation)

In Antarctica U.S. Navy Admiral Richard E. Byrd tested a
specially designed snow cruiser, one of the forerunners
of today's snowmobiles. *(National Archives)*

The USS *Bear* was one of Admiral Richard Byrd's Antarctic supply ships, shown here at the Bay of Whales. Byrd led five expeditions to Antarctica, his first a flight over the South Pole in 1929 (the world's first successful flight there), his last the Navy's Operation Deep Freeze in 1955. About a thousand scientists now work at thirty Antarctic stations during the polar winter, two or three times that many in summer. Antarctic ice, which is as much as three miles thick, is formed from millions of years of snowfall; present day precipitation amounts to only an inch or two. *(National Archives)*

Tidewater glacier, Glacier Bay.
In southeast Alaska ice born of snow that falls on high peaks
reaches to tidewater and calves off icebergs.
(Ruth Kirk)

A 1960s ranger's wife at an outlying station in Glacier Bay National Park harvested ice from bergs stranded on the lowtide beach. She used the ice chunks in a refrigerator pit dug far enough away from the tent to reduce the danger of attack by brown bears. *(Ruth Kirk)*

The signatures of former glaciers include trees buried thousands of years ago by gravel from receding ice (*above,* Muir Inlet, Glacier Bay, Alaska) and scratches cut into bedrock by stones and grit held at the base of an advancing glacier (*below,* Crater Lake, Oregon). *(Ruth Kirk)*

forage for food. Zoologists believe that penguins, which are flightless, evolved from more conventional birds perhaps as long as one hundred million years ago when Antarctica was part of Gondwanaland. The outlines of their behavior were sketched in during that gentler time, and as living conditions worsened penguins adapted to meet them. Now they are locked into established patterns. Any variance means death, given the harsh conditions of the present Antarctic.

Recognition of the peculiarities of emperor penguins, which may be the world's most primitive birds, led to a thirty-six-day winter search for them during the second Scott expedition. This was the part of his polar experience that Cherry-Garrard specifically called "the worst journey in the world." Bill Wilson, who headed the effort, suspected that emperors must nest in winter. In accomplishing their quest the men endured temperatures as low as −77° F., and they suffered pitifully. Yet for penguins the cold and the blasting of the accompanying wind and drifting snow are routine. They endure such conditions for weeks at a time and do it without shelter and without food. The emperor penguins, largest of all penguin species, stand more than three feet tall and weigh up to eighty or ninety pounds. They feed mostly on fish and squid, diving as deep as 885 feet, which is more than double the deepest dive recorded for a scuba-equipped human. Furthermore, they can stay under water for eighteen minutes at a time.

To breed, these birds leave the ocean and march to their established rookery sites on the ice close to land. They arrive in late autumn, always returning to the same place. As many as one hundred thousand may congregate, their weight amounting to thousands of tons. Since the ice isn't stable this late in the year, chunks sometimes break under the weight of the birds and carry great numbers of them to sea. If it merely sags instead of breaking the penguins shift location a bit to save themselves from being flooded by seawater. Selecting a suitable situation on the ice is crucial. The rookery needs to be where the bond between new sea ice and land will be secure, so that winter storms won't set the colony adrift. At the same time it should be where leads will open in summer. And it can't be too desperately far from open water through the winter or the adult birds can't make the incredibly long feeding journeys that are needed while brooding their young. Even at best, the extent of the ice increases through the winter, and these treks commonly become fifty to one hundred miles long. They can't be longer than that or the parents haven't time to walk to the ocean, feed, and return before their young starve.

Courtship begins soon after the penguins arrive at the selected nesting site, and sometime during May or early June, which is the onset of the southern winter, the females each lay a single egg. By this time the journey to the rookery, mating, and laying already have kept the birds two months without food, and they have lost a quarter of their body weight. The females feel an urge to feed, and they need the males to take over nesting responsibilities. To achieve this each female begins to parade with her egg on her feet, snugly held against her warm belly. The male notices and follows. The female then gives a wag of her head, spreads her feet apart, and melodramatically rolls the egg onto the ice. Seeing it, the male uses his bill to hoist it onto his feet against his brood patch. He vocalizes a bit, tries to follow the female as she walks through the rookery, gives up, and settles down to incubate the egg. The female heads for the sea, walking and tobogganing on her belly.

For the male a nine-week vigil begins, perhaps the greatest regularly occurring survival feat in all of nature. Egg temperature must be kept at about 88° F. regardless of the buffeting experienced by the parent bird. To make this possible the lower abdomen of the male has a special flap of skin richly supplied with blood vessels, which covers the egg and holds it fairly firmly in place. He can shuffle about and even toboggan without losing the egg or crushing it, but he can do nothing to escape the tempestuous weather. He can protect his developing embryo but not himself. The only defense for the males is to huddle close together and, peaceable birds, they endure this massing. It gives a rookery the look of an endless series of backs hunched as a living shield over the precious eggs. When penguins on the outer ring get too cold they shoulder into the huddle, and others take their turn at bearing the full impact of the weather.

A layer of blubber initially an inch and a half thick provides built-in nourishment for the male penguin during his ordeal, and he eats snow for water. His feathers insulate so effectively that if the temperature chances to warm anywhere close to freezing, he starts to suffer heat stress. He then fluffs his curious stiff feathers to release the dead air next to his skin and let body heat escape. Each feather is curved at the tip to almost interlock with an adjoining feather, and each has a tuft of down at the base. As many as seventy to the square inch, the feathers ordinarily overlap like tiles on a roof and ward off cold by holding life's warmth within.

In July, the time of the worst blizzards, chicks begin to hatch. Male birds by then have lost nearly half of their weight but haven't yet completed their duty. They must continue brooding until the females return with full gullets to feed the hatchlings and take over warming them. In an emergency the males can secrete a whitish mucus for the young which is rather like pigeon milk, but having to produce it drastically diminishes their own reserves. Usually, the timing is such that the female returns soon after a pair's lone egg hatches, and the male's milk isn't needed. She brings about seven pounds of fish and krill, which is enough for the chick while the male at last ends his fast and leaves her with the brooding. He starts for the ocean to feed with two strikes already against him: He has starved himself for four months and, while this is happening, the ice has widened. His return to the sea is a longer journey than his trek inland from it. Yet despite the seeming odds, the system works. The male succeeds in reaching the water, feeding, and returning to the rookery.

Chicks weigh less than one pound when they hatch but by the time they fledge at the end of summer they have gained thirty-five or forty pounds. Through all these weeks their parents take turns caring for them, and here again the advantage of winter breeding is apparent. As increasing summer warmth lessens the distance to open water, the parents can accomplish their fetching and carrying with greater and greater ease, even though eight or ten pounds of food are needed for a meal by the time the young reach adolescence. The adults are faithful parents. Every mature emperor penguin seems to covet a chick, so much so that a high mortality results from overattention. A chick that strays from its parents' feet may be trampled as six or seven adults that have lost their own chicks bear down upon it. Some chicks get directly loved to death; others crawl under a chunk of ice to escape and end up freezing to death. Even dead chicks are given care. Adults carry them around until their down wears thin. The same impulse during incubation prompts any adult that has lost or broken its egg to snatch a neighbor's egg if it can find one unguarded. That failing, it may work a chunk of ice onto its feet and brood it until it melts. No more than one chick from every ten to twelve eggs laid survives to venture to the sea, but an emperor penguin that lives this long is likely to survive another thirty to thirty-five years.

Adélie penguins are much smaller than emperors. They weigh only about sixteen pounds. Of all the eighteen species and subspecies of penguins, they are the ones most widely known. Unlike emperors, the Adélies breed in summer but even so their lot is a frigid one. Avalanches occasionally sweep into the rookeries, burying eggs, chicks, and adults; or adults may escape from the oncoming snow and thereby doom their young to freezing even if they aren't buried. Blizzards are almost certain to blanket nesting colonies, despite the summer season, leaving the nests as mere dimples in the overall white, each with a pair of eyes at the bottom. In this situation heat from the penguins' bodies melts the surrounding snow and when it refreezes each is walled by ice. Even worse, when this ice melts the nests often flood. Melt streams from nearby snowbanks also threaten submergence at times. To stave off such disaster, brooding Adélies arrange stones beneath their feet to lift egg or chick and their own warm bodies above the water. Stones are treasured inordinately. Neighbors steal them from one another and, curiously, penguins carrying stolen stones act differently than those with stones honestly gained. As if to hide guilt, a thief caught by a rightful owner will instantly drop the stone it is carrying and begin a phony pecking at anything handy.

The Adélies, as the emperors, make long trips from the ocean to their rookeries and vice versa. Their stride is only four inches; their rate of walking about 120 steps per minute. Divide these figures into a distance of thirty or forty miles and the Adélies' odyssey, too, becomes incredibly long. Returning from one of its feeding journeys, a parent bird gives a loud call as it approaches its former nest site, and its chick responds by dashing over to be fed. Instead of immediately obliging, however, the adult usually runs off with its own chick and often a coterie of others in pursuit. During this flight the adult pauses to turn and viciously peck any extra followers and to regurgitate food into the gullet of its own offspring. Only vigorous chicks survive such rites of infanthood, and even they at times are forced to share feedings with skuas or sheathbirds. These scavengers fly boldly at the head of a regurgitating parent, causing it to draw back to strike. This spills some of the krill, which the intruder then picks off the snow.

Once mature enough to take to the sea and feed independently, young Adélies may fall victim to leopard seals and killer whales. As defense their pectoral muscles grow strong enough to give them bursts of speed as great as thirty knots, and they can leap up to twelve feet into the air to escape onto an ice shelf. Man, a newcomer in the Antarctic, is their only other enemy. His helicopters and snowmobiles scare penguins from their nests, leaving eggs and young vulnerable not only to cold but also to opportunistic skuas, which move in quickly. His buildings create lees where such large snowdrifts accumulate that no amount of rearranging stones can ward off ruinous flooding of nests. More directly, men kill penguins to feed to their dogs. In early whaling days, thousands of the birds' fat bodies were burned as fuel while rendering oil from blubber. Some breeding colonies seem not to have recovered yet from these depredations, and there is concern that man's predation may overwhelm penguins and eliminate them as a life-form.

In the Arctic penguins quite certainly would be too vulnerable, not just to man but to polar bears and arctic foxes. Eight species of birds stay in the high north latitudes year round, but all are capable of flight and thereby relatively safe from land carnivores. Penguins, on the other hand, would be defenseless. In the Antarctic they are virtually the only form of vertebrate life ashore for any length of time. The numbers and variety of arctic land vertebrates of course fall far short of those belonging to temperate latitudes; there is little jostling for space, and ecological webs tend to be relatively simple. Nonetheless, arctic life is abundant compared to that in the sterile Antarctic. Lemmings and voles by the million and separate caribou herds numbering in the hundreds of thousands thrive. So do musk ox, bears, wolves, wolverines, and foxes. Each summer the low-elevation, ice-free, ocean-moderated lands so resemble temperate regions that geese and ducks flock there from far away to nest. They have no particular adaptation beyond this ability to travel distances, and they need none. The Arctic is Eden, in season.

Through the endless days of summer, birds and mammals search for food practically nonstop, many of them needing only two or three hours of sleep out of each twenty-four. All need to store part of summer's abundance either in food caches beneath the snow or as layers of fat on their own bodies. To do this arctic species forgo choosiness. Snowy owls, which hunt live rodents during times of plenty, settle for carrion if need be. Purple sandpipers at times give up the search for larvae and eat plants. Arctic foxes, normally sustained by rodents and birds, may instead feed on mollusks. Reindeer nibble seaweed if deprived of lichens and may even devour lemmings. If polar bears can't find seals they turn to fish, birds' eggs, starfish, dead whales washed ashore, grass, or sedge.

The greater the latitude within the Arctic, the fewer the species equipped to cope with winter. Most northerly of those that can are polar bears and arctic foxes. The bears, as the penguins of the Antarctic, are conspicuous on the ice, yet actually depend on the ocean for sustenance. The foxes, in their northernmost range, depend on the bears and hence are also beholden to the ocean. The character of water doesn't fluctuate as that of air does. Its moisture is constant, its temperature relatively so, and no wind blows within its deeps. Diving in it penguins and polar bears lose the insulation of their feathers and fur as pressure drives out entrapped air. Yet water penetrating practically to the skin poses no problem that a layer of fat and metabolic heat can't answer. On the ice surface and exposed to the air, however, plumage and pelage are indispensable as shields against the cold. The body temperatures of polar mammals and birds stay about the same as those of mammals and birds elsewhere. Feathers and fur and fat simply hold in warmth with remarkable effectiveness.

An early written mention of polar bears is in the log of John Davis, an English explorer who in 1585 sailed into Cumberland Sound, on the southeast coast of Baffinland, and wrote: ". . . we espied four white bears at the foot of the mount [and] we, supposing them to be goats or wolves, manned our boats and went toward them; but when we came near shore we found them to be white bears of monstrous bigness." Even earlier records tell of Norsemen trading iron to Greenland Eskimos in exchange for polar-bear skins to use at cathedral altars for keeping priests' feet warm; and a ninth-century note mentions twelve live cubs shipped from Iceland to Norway.

The bears range as far as 80° north latitude, never far from drift ice and in fact preferring the region of the greatest ice movement. They seldom wander more than a mile inland, and they avoid sheltered inlets where the ice is relatively stationary. Huge animals up to ten feet long and seven hundred to one thousand pounds in weight, polar bears nonetheless are extremely agile. They lope along at twenty to twenty-five miles an hour, easily leap ice ridges five to six feet high, and jump from lookout posts as much as twelve feet high. Even on polished ice they move confidently, gaining traction from close-set stiff hairs on the soles of their feet. In soft snow their ten- to twelve-inch paws act as snowshoes and, when the surface won't support their weight, their long and powerful legs still let them travel through snow better than any other arctic animal. On thin ice they spread their legs until their bellies drag and in this position slither along for considerable distances. As swimmers they are nearly as successful as seals. They can travel four or five miles an hour and stay under for two minutes, eyes open, ears and nostrils closed. Their paws provide paddles, and if they choose they can swim porpoise-style, leaping ten to twelve feet out of the water at a time. Their leg bones are filled with buoyant oil rather than marrow; their skin is spongy and oily; their fur is oily enough to be waterproof for short periods at negligible depths.

The territory of polar bears is among the pressure ridges and leads of the pack ice where they can climb up to look for seals. It is a noisy realm, so much so that the sound of rifle shots holds no great terror. In *The Friendly Arctic*, Stefansson wrote: "The floes buckle and bend, and are thrust up into great ridges; then tilt, snap, and crash over, groaning and booming like surf cannonading on a rocky shore." In such a setting the crack of gunshots scarcely is startling and a keen sense of hearing has little value. Eyesight is not highly developed in polar bears either. Their eyes and optic nerves are small and unless something is moving, they don't notice it at distances of six hundred feet or more. It is smell that the bears most rely on. As many as thirty-five of them have been reported—and shot—gathered about a carcass to feed. The number is surprising because odors don't carry well in the chill air of the Arctic, and polar bears by nature are solitary hunters. For so many to congregate suggests an ability to perceive carrion from impressively great distances.

They need this sense of smell in hunting, for most of their kills are seals taken at breathing holes. Snowed over, the holes are not apparent to the eye, but bears (and dogs) can detect them by smell. A hunting bear abruptly changes direction to angle off directly to a hole as soon as the odor of seal is detected. The olfactory clues alone induce taking up vigils that often grow lengthy. Eskimo hunters remain motionless by a hole as long as three days and nights if need be, and polar bears seem to do the same. Their beds are found worn deep into the snow. Before settling down to wait, however, a bear scoops out the seal's breathing hole enough to reach in easily with its paw. Then it refills the crater so the approaching seal will see nothing amiss. Next the bear lies down with its nose buried in the snow, seemingly to diffuse its breath so that the seal will have no warning of a predator's presence, or perhaps to hide the black dot of the nose which flaws otherwise perfect camouflage. Seals stay below water only seven or eight minutes at a time, or up to twenty minutes in an emergency, but they each have nearly a dozen holes so no particular one is sure to be visited very often. When a seal does come to a hole it lets out air in a bubbly gasp—and the waiting polar bear pounces with claws and teeth and pulls the prey onto the ice.

Sometimes two bears hunt together, one waiting by a breathing hole while the other stops up all the others nearby, thus raising the odds of a seal's returning to the hole where the first bear waits. Or both bears may walk almost carelessly to a selected hole; then one leaves, letting any lurking seal think danger has passed. Another technique is to lure a seal to a hole by scratching and patting at the snow in imitation of seals lying out on a summer ice floe. Eskimo hunters use special scratchers of ivory for this. Polar bears create the effect with their claws.

Bears also hunt seal pups in their dens beneath three or four feet of snow. These chambers are hollowed out by females at the surface of the ice for use in giving birth and caring for their young. A bear locates a pup by smell, then rears up and strikes at the roof of the den with its forelegs. If need be it will back up and run and jump several times to break through a strong crust. Once the hole is made, the bear reaches in with a paw and hooks out the pup. Adult seals, too, occasionally are captured out of water. In fact, polar bears stalk with impressive stealth and effectiveness. In winter their clean white coats blend well with fresh snow, and in summer their soiled yellowish coats match the hue of old ice. Creeping catlike, a bear lowers onto its belly, pulling along with its forelegs and letting its hind legs drag. Or it may push with its hind legs, holding its forelegs tucked under its chest like sled runners. Sometimes a bear even pushes a chunk of ice ahead with its nose as a screen. By these means it glides to within twelve or fifteen feet of its prey, then lunges. Or, while still some distance away, the bear may choose to slip into the water by backing to the edge of an ice floe, putting in one hind leg, then the other, holding onto the ice with its forepaws all the while. Once in the water it submerges all but its snout and swims practically without a ripple until close to its quarry. Then it dives and shoots up out of the water, or even through thin ice, and lands on the seal and crushes its skull with a single well-aimed stroke.

About every three years she-bears give birth, usually to a single cub but occasionally to twins. For this occasion many females leave the drifting pack ice in fall and go ashore to find a den. Others seek out hollows within pressure ridges. Either way, a bear searches for some time, prospecting for just the right snow conditions. In parts of the Arctic the females that go ashore travel into the mountains to where snow drifts form early on ice slopes. In forested country they may den beneath the undercut vegetation of a riverbank or in the cavity left by upturned tree roots. Once a suitable hollow in old snow is found, the bear enlarges and shapes it to answer her needs. The architecture includes a distinct entry passage sloping upward to prevent the den's warm air from draining out. Above the entry she excavates an oval chamber six or seven feet long and about four feet wide and high. Body heat quickly warms this cavity to forty degrees or more above the outside temperature, and the heavier the snow on top, the better the insulation and the warmer it will be. The bear's breath keeps a small ventilation hole melted open (and Eskimo hunters rely on their dogs to sniff these out, just as they use dogs to locate seals' breathing holes).

Polar bears don't really hibernate. The female lives on fat reserves for four or five months and swallows mouthfuls of snow for water. She wakes to give birth to her young, which weigh only one or two pounds, and to suckle. She also rouses to full activity if an avalanche damages her sleeping chamber or, when denned on the pack ice, she wakes if a berg starts to tip and threatens disaster.

Adult male bears, barren females, and yearlings may or may not den for the winter. If they do, it seldom is for long unless the temperature drops exceptionally low, or seals are particularly difficult to hunt. During blizzards all bears seek shelter. Once in a while they even enter Eskimo igloos. More commonly if there is no shelter, they pivot round and round in the snow to gouge out a hollow large enough to curl into, and then they sleep.

Polar bears once were believed to drift haplessly with the ice, rotating clockwise with the current. This now is known to be untrue. Males and females find one another at traditional breeding grounds on a regular seasonal basis, and dens are used almost every year. The geographical rhythms of their lives thus are not at the mercy of drift ice, but bears nonetheless may float great distances with the pack. They have been found on floes two hundred to three hundred miles from the nearest land, and a polar bear tagged in Svalbard a few years ago turned up a year and a half later in southern Greenland, a distance of two thousand miles. Some bears pass through the Bering Strait on the ice and drift down the Alaskan coast to Nome. In the past they drifted as far as the mouth of the Yukon River. On the Asian side of the Pacific one occasionally still wanders to central Kamchatka, and polar bears previously were known along the coast of Hokkaido. They are reported there in Japanese imperial records as early as A.D. 658.

How many polar bears remain in the Arctic currently isn't known accurately. Estimates are around ten thousand to twelve thousand, and their number no longer is dwindling thanks to the treaty that resulted from the polar bear conference held in Oslo in 1973. In it Canada, Denmark, Norway, Russia, and the United States agreed to prohibit the killing or capturing of polar bears over what amounts to about 95 percent of their range. The use of aircraft and ships is specifically banned in hunting bears; snowmobiles are allowed only in Canada and there only when used by natives engaged in subsistence hunting. No hunter in any category may travel more than one day onto the ice of the Arctic Ocean to hunt. Denning areas are given special protection. Before the treaty, survival had been in jeopardy. For no matter how superb polar bears' adaptation to the ages-old rigors of the Arctic, nothing but man's own decision can save them from his capacity to kill. Bears' masterful hunting can fill their bellies and fuel their bodies. Thick fur and canny denning can shield them from the bitter cold of the long polar night. But no defense is sufficient against the determination of human commercial and sport hunters except, hopefully, the treaty.

Arctic foxes, the only other animal to feed as far north as polar bears, have escaped profound threat to continued survival. Their range extends to every tundra coast and island of the Arctic, and in winter they commonly travel the sea ice with bears. Inhabiting such a variety of ecological niches gives them an advantage as a species, yet their fur is the thickest in proportion to size of any mammal in the Arctic and is prized accordingly. Eskimos in Canada alone presently market from ten thousand to eighty thousand pelts per year, each valued at around thirty dollars. For trappers the arctic fox is akin to a cash crop, and money is the new Eskimo open-sesame for everything from snowmobiles and rifles to boxes of cake mix and packets of steel needles.

Arctic foxes have an additional advantage in being adaptable in their food habits. They stand out even in a region where catholic taste is requisite to survival. In many parts of the north, lemmings constitute such a staple in the diet that vixens successfully raise full litters of ten or twelve kits in years of abundant supply, and perhaps even have a second litter. When lemming numbers are low, the numbers of foxes also drop, and those that survive turn to other foods. In other parts of the Arctic, bird rookeries provide the major food source. A trapper reports finding a fox's food cache beneath eighteen inches of snow with thirty-six baby auks in it, each with its head bitten off, the tiny carcasses all neatly arranged side by side and pointing in the same direction. With them were the bodies of four snow buntings, two immature guillemots, and a quantity of auk eggs. The supply would be enough food for a month, and in normal years each fox has several such caches. But when food is in short supply, or caches are sealed beneath hard-crusted snow, matters are different. At those times particularly, but also in normal years, coastal arctic foxes follow polar bears on their seal hunts, and inland foxes follow wolves which, in turn, follow musk ox and caribou. Nansen reported polar bears accompanied by arctic foxes on floes 250 miles north of the New Siberian Islands, and Peary found fresh fox tracks 300 miles north of the Greenland coast. Desperate for food—and curious by nature—foxes even have been known to board ships frozen into the winter ice.

The relation of arctic foxes with polar bears is one-sided. The foxes benefit from it; the bears are unaffected one way or another. A fox will follow polar bear tracks, then curl into the snow and sleep until a seal has been caught. Often a raven perches nearby, also waiting. When the bear makes its kill, it usually skins the seal by biting around snout and flippers and jerking off the hide to expose the rich blubber. This the bear feasts upon, then it ambles off, leaving the seal carcass to be scavenged by the waiting fox and the raven.

Feeding thus without spending much energy in hunting is an essential part of survival for arctic foxes. Their incredibly rich fur is another part. Long outer hairs lie as a shield over woolly inner hairs, seemingly held in position by opposite electrical charges between the two layers. Legs, neck, and tail are short, which minimizes circulatory heat demands. Ears and nose are small and furry. The combination lets foxes survive even in the high Arctic without shelter and without raising body temperature higher than that of foxes anywhere else, about 100° F. If it weren't for this remarkable fur and compact body shape, an unsheltered arctic fox would have to generate about ten times the body heat of a desert counterpart. As it is, even when the air drops more than 150° F. below body temperature, these northern foxes are comfortable. In comparison, an unclothed man starts to shiver and thereby increases metabolic heat production when the air temperature falls about thirty-five degrees below his body-core temperature.

Of all the species living in the Arctic and the Antarctic, man is the most vulnerable. He stands tall and receives the full buffeting of the wind. His body has no natural insulation or water repellency. He can't cosset himself from the bite of the winter air by dropping into torpor or diving into the constancy and relative warmth of the ocean. Clothing and shelter are essential for survival. Yet, though he lacks innate adaptation to the polar realms, man can dictate their ecological future.

CHAPTER

5

SHELTER AND CLOTHING

THE WHITE EXPANSES OF THE POLAR REALMS REMAIN TODAY, the blizzards howl, and the cycle of light and dark continues through the months first one way, then the other. The land hasn't changed, but man's position within it has. Nansen's crew watching *Fram* freeze into the ice off Franz Josef Land, or Scott's men wintering at Hut Point, scarcely could be expected to empathize with the life conditions now experienced by their successors.

None of the polar pioneers would stand incredulous before the changes, of course. They were too much men of vision to permit such disbelief, and they worked too hard to improve the ways of coping with what is perhaps the most forbidding environment on earth. Yet how could they dream of trees grown on the arctic coast purposely to lift morale, or of men running naked in Antarctica to break the monotony as an expression of the blithe "streaking" fad that flared into fashion in the early 1970s. At the American polar research station the exclusive "200 Club" limited its membership to those men who sat for fifteen or twenty minutes in a sauna, then donned only boots and dashed 150 feet to the exact spot of the Pole for a photograph. Requirements stipulated that the temperature difference between the steam bath and the outside air must be at least 200 degrees. Otherwise membership was denied, although applicants were free to try again.

The home base for this particular elite club was the New Pole Station, completed in 1974 at a cost of six million dollars. Four metal tunnelways like gigantic corrugated sewer pipes 44 feet high and 180 feet long house a generating plant, a depot with 150,000 gallons of fuel stored for winter use, a maintenance shop, and a biomedical research laboratory. A separate geodesic dome 164 feet wide and 50 feet high shelters three additional prefabricated buildings from such climatic realities as temperatures potentially lower than −100° F. and snow loads of 120 pounds per square foot of roof surface. Included in these buildings are living quarters, scientific work space, a library, a communications center, and a post office. This new marvel of a station, one quarter mile from the Pole, replaces the nearby Old Pole Station which now lies covered with forty feet of snow. The old buildings date from 1957 and were intended for three years' service, yet actually continued in use until 1975. By then they had become mounded over with new snowfall and crushed by ice. The ice formed as heat from the buildings melted the snow surrounding them, only to have it refreeze and thereby add to the pressure against walls and roofs as it expanded. Cracks resulted, and they increased the escape of heat and added still more to the icy straitjackets holding the buildings. The new design should avoid the whole destructive sequence. Cold air blown into the shell's interior comes into contact with the buildings, but no snow can reach them, and no ice is expected to form. Insulation beneath the entire structure should prevent the buildings from sinking into the snow.

Much the same thirst for knowledge that activated the early explorers prompted establishment of this new base and the others in Antarctica. Scientists from a dozen nations and scores of academic disciplines today work together toward a common goal of understanding. In the North the situation differs somewhat. On the arctic slope of Alaska and Canada's Northwest Territories, hunger for oil and gas gives the thrust for current development, and instead of governments and scientific institutions financing the undertakings, costs are borne by companies from Arco and Exxon and British Petroleum to Great Canadian Oil Sands and Syncrude Canada Ltd. At Prudhoe Bay on the Beaufort Sea company managers and staff personnel of British Petroleum, Ltd., live and work in a veritable space station set on steel pilings and designed to provide "a variety of living experiences," as one of the architects expresses it. The most luxurious of all North Slope camps, rooms here are carpeted, walls brightly decorated, food from steak to fresh fruit is served around-the-clock, and a glassed-in Astroturf courtyard offers a place for jogging or volleyball, or for looking nostalgically at the trees of a small arboretum. An English artist who was commissioned to paint a scenic wall mural instead painted bright stripes and zigzags of saturated blues and reds and greens onto the wall of the mess hall. The scenery was too drab, he said, and would only demoralize the men if brought inside. Workers swell their bank accounts and let a certain macho pride compensate for aching muscles and boredom. Such is today's life in the snowy north that one evening at an Arco subcontractor's camp at Prudhoe Bay, while waiting for a phone, I heard a boy of nineteen or so saying into the mouthpiece, "Yeah, but I've never made twelve hundred dollars a week before in my whole life."

This instant wealth carries over to the booming communities that serve the camps, and speculators seek their fortunes by the usual dubious schemes while hookers cruise the streets in motor homes picking up clients, a thoroughly modern touch. Aspects of the old Klondike gold rush are reborn and updated. How all the bonanza of black gold and gas will get to market remains another matter, as does what it may do to the environment en route. But meanwhile planes and helicopters crisscross the skies from Prudhoe Bay to the Mackenzie delta every possible minute, and leaders in the American union-labor movement, ironically, dart to and fro in remarkably efficient Japanese-built turboprop planes. In Eskimo and Indian camps, dogs howl their timeless compliments back and forth, and on the tundra biologists equipped with mountains of gear track and photograph and analyze every creature that walks or flies, from musk oxen and caribou to ptarmigan and courting sandhill cranes. Hours of preparation precede each move; then plans get canceled, or the weather closes in and disorganization reigns. "But, Lord! The money going through the mill here," as one Canadian observer at Inuvik puts it. "Ask and it shall be granted."

As early as the sixteenth century, men began to approach scientifically the problems of living in the Arctic. True to their era, they ignored the ways in which native peoples already had found solutions, and they sought through their own ingenuity, diligence, and patience to render the Far North habitable, or at least to lessen the energy demanded for mere survival. A Russian manuscript dated 1557 deals with geocryology (the study of frozen earth), rightly attributing permafrost to climate and specifically to the balance between frigid air and heat loss from the earth. Soviet researchers today continue the drive to master their north through investigation that is partly theoretical, partly applied and long since worked out in meticulous detail. For example, in the northern Ural Mountains, where winter lasts half the year, a recent study of snowload on unheated industrial buildings disclosed that clerestory windows tend to accumulate an undue amount of snow, especially on the lee, or shaded, sides of the buildings. Consequently such windows are ill advised, and if they must be used they should be M-shaped rather than rectangular. Similarly, cities throughout the Soviet north are planned for maximum protection from drifting snow. This involves careful attention to siting, along with designing buildings generally tall and smooth sided, laying out main streets parallel to the prevailing wind, and raising roads outside the settlements themselves onto earthen embankments.

Much the same sort of work has been done in Canada where the National Building Code has been revised to incorporate data resulting from laboratory analysis of replicated conditions. In one study, model buildings, fences, and trees scaled at one inch for every sixteen feet were set up in a twenty-three-foot water flume, and storms were simulated with white sand carried by water currents to represent windblown snow. Engineers could observe "snow" passing over the model buildings and settling around them; they could compare the loads accumulating on roofs of various pitches and textures with those accumulating at ground level, and could study the interaction of various "wind" speeds and note the effect of obstacles. Missing were the physical properties of true snowflakes and air, and the effects of sunshine and of heat loss from buildings and snowpack, but even so the results from the water flume matched theoretical data from wind tunnels, as well as hundreds of aerial observations of actual conditions.

In the arctic hinterland of North America and the Soviet Union, in Greenland and Iceland and in Antarctica, additional attention focuses on snow and ice not as a construction problem but as a building material. The logistics of bringing materials into these regions constitutes an endless and costly dilemma and, as with any remote undertaking, meeting needs with local supplies is far preferable to hauling in goods from the outside. In polar regions, this means building with snow and ice. Consequently, in the winter of 1950–51 the United States Navy started experimenting at Point Barrow with ways of artificially thickening natural ice. They diked an area 150 by 400 feet and repeatedly flooded it with seawater, letting the thickness of the ice increase a layer at a time from 6 to 16 feet. Eventually they learned that the strongest ice is produced by supplying the water either in a fan-shaped spray or with a high arc. Having developed a production method, the artificial ice was put to use the following season to make a half-mile runway. Two problems wrecked the undertaking, however. Large unfrozen zones caused serious weakening, and pressures within the surrounding, natural ice caused roughening of the artificial ice. The failure sent the ice engineers back to their dikes, and this time they came up with a circular sheet of ice, which proved less susceptible to surrounding pressures.

Only a few years earlier, the military needs of World War II had brought the first concentrated determination to use snow and ice instead of cursing them. At the time few aircraft had skis, and compacted snow alone didn't offer the runway strength needed for wheeled landings. To solve the problem ordinary rotary tillers intended for farm use were freighted north; so were various rollers and drags. And men began learning to build with snow. They tried mounting a fuel-oil burner on skis in the hope that heating snow while mixing it would make a better building material. To break the natural layering of the snow and mix grains of various sizes and temperatures, they welded special tines onto blades. That way the union of crystals one to another should be sped. In Greenland the U.S. Army experimented with a machine that whirred blades as fast as three hundred rpm and churned a nine-foot swath of snow to a depth of five feet. It blew snow up and over and settled it back down again by a special chute. A second monster traveled behind to level and smooth the surface before the snow could solidify, and after it came a roller to compact the whole. Another Army test centered on melting snow, then heating the water and spraying it through nozzles originally meant for use with bitumen. The effort proved more attractive in theory than in application, unfortunately, because it took too much fuel to heat the snowmelt enough to keep it from freezing in the nozzles and clogging them hopelessly instead of spraying out.

The Navy, too, set about developing a means of processing snow. They used a standard engine-driven earth pulverizer fitted onto skis and followed with a drag to smooth berms. A single pass with the equipment mixed and blended the snow particles but left them as elliptical crystals one to five millimeters long, a shape and coarseness that prevents close contacts between grains. But by making three passes over the snow and allowing for age hardening, the pulverizer and drag produced a surface six times as hard as untreated snow. Sawdust an inch deep protected it from gasoline and oil spills, the sun's rays, temperatures above 25° F., and wear. When the Winter Olympics were held in Squaw Valley, California, in 1960 this technique developed by the Navy was used to solve the problem of parking lots. A forty-acre field of depth-processed, surface-hardened snow covered with sawdust successfully accommodated three thousand cars on opening day, and a Navy team kept on processing more snow until they had provided space for nearly four times that many cars. Access to the parking lots was by a mile-long road built from twenty-five hundred cubic yards of snow specially trucked in and packed into place. At night maintenance crews patrolled the road and the lots to brush sawdust onto bare spots and to press a mixture of sawdust, snow, and water into wheel ruts.

A technology wholly separate from constructing parking lots and runways by compacting snow involves reinforcing snow with fiber for use as walls and beams. The process is akin to reinforcing concrete with steel, or plastic with fiber glass. A host of materials have been tried: wood fiber, asbestos, strips of paper, newspaper mash, jute cordage, shredded excelsior, strands of fiber glass, and metal mesh. These are intended to counteract the natural tendency of snow to creep and to overcome its inherently low tensile strength. Snow mixed with various fibers is blown directly into forms, then wrapped in parachute covers and allowed to age-harden for four months. Initially beams were cast of snow alone, without reinforcement, to compare against the others. All these broke as the forms were being stripped off. So did beams with jute cordage, which tended to pull out without any bonding action at all. Others tested well. A snow beam reinforced with metal mesh didn't break even when supported only at each end and subjected to two men jumping up and down at its center—an unsophisticated test but the best that could be devised since no conventional load-test facilities were at hand. Excelsior proved effective as a strengthener. Wood fibers did not. Even when mixed thoroughly into the snow they tended to settle to the bottom before freezing was complete. Asbestos fibers had the opposite problem. They floated to the top, probably borne by air bubbles, and pouring the snow a layer at a time didn't eliminate the problem because the fibers bridged poorly from one layer to the next.

In U.S. Army experiments in Greenland an artificial aggregate has been developed by mixing silicates and aluminosilicates with snow to form "permacrete." This can be handled almost like ordinary concrete, providing the temperature stays below freezing. Permacrete columns now line tunnel walls and support roofs, and permacrete bricks have been laid into masonry walls. Chairs have been poured, platforms built, and reservoirs for nonfreezing liquids constructed, all using permacrete. Stirred into a slurry with pigmented clay, the snow mixture even serves as paint.

Probably the origin of building with snow lies somewhere back in the white dawn of arctic prehistory, and peoples living at high elevations elsewhere also may well have built with snow. It is Eskimo snow technology that is best known and seems to be man's most highly developed, at least for the days before permacrete. Igloos of snow blocks have become a stereotype, although they were only one part of Eskimo snow use. Structures built of snow formerly included high platforms to keep meat out of dogs' reach and windbreaks to shield igloos from blizzards and to direct the inevitable snowdrifts to where they were the least bothersome. Hunters waiting at seals' breathing holes put up low snow walls to give protection from wind; for target practice some used snowmen. Fathers made toy animals out of snow for their children, and on days that weren't too cold Eskimo women cleaned their hands and knives with snow after butchering animals from the hunt. On extremely cold days—which are frequent—snow freezes tissue and can't be used on bare hands. If more water was needed in a simmering stewpot a housewife needed only to scrape snow from the igloo wall behind the blubber lamp, and add it. Or if the roof began to drip she could use small pieces of snow as blotters.

With few exceptions Caucasians entering the polar world have hesitated to copy the arctic natives, either out of ignorance or a woefully misguided feeling that so-called primitive ways have nothing to offer modern men. It is a snobbery that has caused suffering and death, although there are a few early accounts that mention native use of snow respectfully. For instance, Shackleton in Antarctica wrote in 1909: "There are no Eskimos . . . whom we could hire, as Peary did, to make snowhouses for us." Similarly, Nansen several years earlier had acknowledged the effectiveness of a well-built snow igloo. The first winter after *Fram* froze into the ice, a specially designed tent had been used for making magnetic observations, as planned. But during the expedition's second winter, "the observer, Hansen, greatly increased the comfort of his work by building, along with Johansen, an observation-hut of snow, not unlike an Eskimo cabin. He found himself very much at his ease in it, with a petroleum lamp hanging from the roof, the light of which, being reflected by the white snow walls made quite a brilliant show. Here he could manipulate his instruments quietly and comfortably, undisturbed by the biting wind outside. He thought it quite warm in there too . . . so that he was able without much inconvenience to adjust his instruments with bare hands."

Preparation for modern expeditions now often includes study of native construction methods. Eskimos and Siberian Koryaks sometimes built houses of willow branches covered with skins and banked with snow too loose to cut into blocks. Sometimes they took heated rocks inside and left them until the snow of the walls began to melt, whereupon the stones were removed. Adapting these techniques, today's military men are instructed in laying a parachute over

a pile of spruce and willow branches—or aircraft fragments—
and covering the whole with a thick layer of snow as insulation. In principle this method depends on sublimation
and recrystallization within the snow, just as in the case of
compacted airplane runways or ski-resort parking lots. It
is the same cementing that occurs in natural snow cover
but at a speeded rate. The process of scooping up snow to
cover the form mixes "warm" snow crystals against "cold"
ones and thereby spurs recrystallization.

If a weather balloon is at hand it makes a form superior
to anything else, according to a U.S. Army instruction manual. Simply excavate a hole and place the inflated balloon
in place so that half of it rests down into the snow. Cover
the upper half with snow a foot deep and let it set for an
hour. Then tunnel into the balloon; deflate it and leave the
snow dome standing free. Complete the shelter with a door
of snow blocks, or clothing, and glaze the interior walls and
ceiling with heat from a stove to increase structural strength.

In four hours the walls will be about ten or twenty times
stronger than they were at first. Another method, suitable in
an emergency, is to dig a trench into a drift and roof it with
snow blocks or a covering of canvas, or whatever is at hand;
then pile on snow. Such trenches seem warmer than domed
huts, probably because the almost infinite thickness of the
walls prevents all possibility of heat loss from wind.

Temperatures recorded for a weather-balloon snow house
with the floor dug through the snow to sod indicate the astonishing effectiveness of such shelter. During one test outside air temperatures dropped to −55° F., yet with the door
sealed two men sleeping inside the snow hut, without heat,
experienced a fairly comfortable 19° above zero. Even when
unoccupied at −28° F. the inside temperature at sleeping
level registered fourteen degrees above zero. With two
candles and two men, the inside temperature after fifteen
minutes stood at +23.5° F. while an outside thermometer
showed −48°. (Wind speeds were not recorded.) Much of
the warmth in these tests came from the floor, derived from
the radiation of the earth. The sod surface outside the shelter
averaged from forty to sixty degrees above the daytime air

temperature. Thermometers there read from +16 to +20° F. while those registering air temperatures showed −12 to −40°.

Such snow houses can be impractical when the weather alternately freezes and thaws, as for example is typical in autumn. Also, when temperatures are as mild as 10 to 20° F. snow houses begin to develop bothersome melt holes in the roof, and walls get so wet it is hard to keep clothing dry. Even at the outset, a very real limitation is the matter of finding proper snow for building. At the optimum season and in ideal terrain, snow won't necessarily be right. If it is newly fallen it will be loose for blocks, and trampling it may well fail to bring it to a consistency that permits cutting (although fresh snow can be mounded onto a form, such as the weather balloon). At the other extreme, if snow is old and wind-packed it may be so hard that getting blocks requires real chopping and hacking and, for all the work, produces blocks heavy to carry and hard to work with. They will be difficult to shape, slippery, and so compact that they readily conduct heat to the outside and result in a cold house.

Sometimes even with "good" snow, tents seem tempting. They can be erected in less time than it takes to build a snow house, although two men working together can cut blocks and fashion a snow shelter in an hour, given favorable conditions. Time gained with a tent in the evening, however, is lost in the morning. You walk out of a snow house and are done, but a tent must be taken down, brushed, folded, and loaded. Also, the breath of sleeping men and the steam from cooking pots condenses on the inside walls of a tent, and hoarfrost showers down like snowfall as the canvas flaps in the wind. Tents do warm more quickly than snow houses—and they lose their heat just as fast once the stove is shut off. A tent will be −10 to −20° F. at an outside temperature of −50° even on a calm night. With wind, it will be worse. On the other hand, Everest climbers learning to build snow shelters in the Canadian Rockies a few years ago reported the inside temperature first stood at −27° F. Then as the men lit a stove to fix tea, "off came fur caps and mukluks and before we realized it we were enjoying all the warmth we could wish for." In twenty minutes the little stove had raised the temperature to +65° F.

Snow houses never were universal among Eskimos. They provided the standard winter dwellings along the central arctic coast of Canada and in parts of Labrador where families wintered on the sea ice to hunt seals. Greenland and northern Alaskan peoples built them when traveling or hunting. The word *igloo* actually means any kind of house, not just one built of snow, and most Eskimos lived in sod houses built with driftwood or whalebone frames and set halfway into the earth, for warmth. In his book *Hunters of the Great North*, Stefansson tells of his introduction to "the real Eskimo snowhouse," although he previously had camped in a shelter of snow blocks set as vertical walls and roofed with skins. The first step was finding the right snow. As night approached Stefansson's Eskimo companion, Ovayuak, and his wife began checking drifts until they found a promising one. Their footprints gave the first sign that the snow was suitable: too hard, and fur mukluks leave no mark; too soft, and they sink in so that the outline of the whole foot shows; just right, and the imprint is barely plain enough that a man's trail can be followed. Ideally a drift should be around four feet deep and uniform. To check, Ovayuak probed with a rod (traditionally of caribou antler), driving it with a steady motion. If it alternated between slipping in easily and being a bit hard to push, the snow would not do; it was too stratified and blocks cut from it would break into layers. But if the probe slid in evenly, the business of cutting blocks began.

For this task Ovayuak used a special snow knife with a caribou-antler blade about fifteen inches long. He cut vertically, producing blocks twenty-five inches long by fourteen inches wide, and any blocks that came out thicker than others were trimmed until all were about four or five inches through. In shallow drifts he cut horizontally, but this is awkward.

The trick in building an igloo is to produce a spiral and to undercut the inner edge of the blocks so that each one tilts slightly inward when set into position. Ovayuak laid the first course of blocks to form a ring ten feet across, and then he tapered the upper edges of three of the blocks into a sharp diagonal. This was to force the second tier to begin rising at an angle, setting a spiral for the entire house. Each successive round was tilted inward slightly more than the preceding one, a key factor in producing a smooth dome as the roof blocks were placed. Ovayuak worked from the inside. His wife busily cut new blocks, and Stefansson handed them in, first over the rising walls and, when that became awkward, by passing them through a hole cut at ground level. Five tiers totaling forty or fifty blocks were needed to complete the house. The final closing of the roof is "easier than anything," Stefansson wrote. "When you get near the roof the circle you are working on is less than half the diameter of the original ground circle. The blocks, therefore, meet at a much sharper angle and you can lean them more squarely so they support each other better." The last block is fitted into a hole trimmed to give a perfect match. The builder "takes up a particular snow block, trims it so it is a little thinner than average, puts it on end and lifts it vertically through the hole, so that if you are outside you can see his two arms sticking up through, holding the block. He now allows the block to take a horizontal position in his hands and lowers it gently down upon the opening so as to cover it like a lid."

With this, the basic house is done except for gently rubbing soft snow into the chinks and shoveling additional snow over the outside to bank three feet deep around the base, thinning to eight or ten inches by the third tier of blocks. The roof is left with only the thickness of the original blocks. The entry is fashioned both from the inside and the outside. As the final block was fitted into the roof, Ovayuak was completely shut inside, for he had filled the hole through which Stefansson earlier had passed blocks. Consequently, "with a shovel his wife now dug a trench about three feet wide . . . [tunneling] under the wall of the house to meet a hole that Ovayuak was digging down through the floor." The trench later was roofed to provide a typical combined porch, entryway, and storage anteroom. Ideally this entry tunnel should include a right-angle turn, to deflect wind. Better yet, it may be T-shaped to provide for closing either doorway, or neither, as the wind changes.

Furnishing a snow house calls only for platforms, also built of snow. A large one, taking up two-thirds of the floor space provides a bed and working and lounging area. A smaller, adjoining platform serves as kitchen table. To build these, rows of blocks are first set on edge to make the fronts of the platforms, and then loose snow and fragments of blocks left from the house construction are filled in behind. This material is packed down hard and smoothed and, if available, wood or bone may be added as an edging to prevent wear from climbing and sitting. A double layer of caribou furs is laid onto the sleeping platform as a lining, the first one placed hairside down, the second hairside up; sometimes an insulating layer of caribou ribs or waterproof kayak covers goes underneath. Blankets or more furs are laid on top as bedding. Other linings are rarely used by the Eskimos, except for the Baffin Island and Igloolik (central Canada) groups who hang skins from the inside of the walls. Reducing the temperature differential between men and icy walls this way theoretically reduces the heat loss radiated from the occupants' bodies. Such liners also lessen glazing, an advantage since glazing decreases the insulation of snow walls. It also strengthens walls, however, and makes them more windproof. Dripping is not a major problem even without wall liners. Water tends to run down the curve of the dome, providing it is free of irregularities, and snow at the base of the walls, which is too low and cold to glaze, blots it up.

A window of freshwater ice about three inches thick used to be set into the wall near the cooking area, sometimes in conjunction with a snow block mounted on the outer wall as a reflector. Freshwater ice is clearer than sea ice and consequently was so much preferred that Eskimos often carried a panel of it from an inland caribou camp to the coast where fresh water, frozen or liquid, is not easy to come by. If heat from cooking grew excessive, the ice window could be curtained temporarily with a skin to prevent its melting.

Occasionally a cluster of snow houses would be built with a common entry to simplify visiting back and forth and sharing routine indoor chores without going outside. Similarly, construction of communal dance halls consisted of roofing over four circled igloos and then knocking out the inner walls. Sometimes single igloos were built thirty feet in diameter and ten or eleven feet high in the center. Because of their dome shape such houses, large or small, have great inherent strength. "During the evening," Stefansson wrote of his first experience in one, "I asked Ovayuak whether there was any danger of the house caving in on us during the night and he laughed at me. When we were about to start [the next morning] this conversation apparently recurred to him, so he asked me if I would like to try how fragile the house was by climbing on top of it. I hesitated a moment, and he ran up on the roof himself and stood on the peak. I then clambered up after him. Had there been ten of us our combined weight would not have broke the house down."

The statement is true—and fortunate—for polar bears weighing half a ton have climbed onto snow igloos as vantage points in a flat land. When first built, the walls are fragile and even chinking with loose snow is done gently lest blocks be broken. But once the snow has age-hardened the dome shape is capable of holding up despite great weight and pressure. Only a sharp blow threatens it so long as temperatures stay low. "The case of an egg is analogous," Stefansson concludes his discussion. "You can easily break it with a sharp blow, but it is not so easy to crush a raw egg by squeezing it in the hand if the pressure is applied uniformly."

For the men of early polar expeditions living in huts and tents, ventilation posed a serious problem, and sometimes it does so today when tents get so ice-coated that their fabric loses its porous nature. On the other hand, in a snow house no one need worry. With a hole in the roof and the entryway left open, fresh air circulates constantly. In fact a standing man may experience all of the world's climates simultaneously: His feet will be in a polar zone, his middle torso temperate, and his shoulders and head tropical. Eskimo children up to five or six years old customarily went naked, and adults often wore only knee-length breeches. "A hot house is good for you," Eskimo people used to maintain. "You can go outside and cool off." Sweating was no bother in the old days. When wood was at hand, men shaved blocks of white spruce into quantities of long, excelsiorlike curls that were piled into corners and used to wipe the sweat from faces and bodies. When white men's towels became available they replaced the shavings—and grew increasingly soiled with no way of washing them, whereas the shavings were used a clean handful at a time, and afterward discarded.

A single blubber lamp traditionally furnished heat for the snow houses. It consisted of a piece of soapstone about a foot across carved as a shallow bowl among the western Eskimos and as a rimmed, flat crescent to the east. Oil filled the depression, and a slightly raised platform held a wick of dry powdered moss, wood sawdust, or fine scrapings of walrus ivory. Seal blubber most commonly furnished fuel, although walrus, beluga whale, and narwhal blubber were preferred in some districts, shark liver oil in others, and haddock oil in an emergency. Inland people got along with caribou or musk-ox tallow, poor substitutes for the rich oil of marine mammals both in quality and quantity. A seal weighing two hundred pounds may yield half its weight in blubber, a luxurious ratio of fuel to food. Except in starvation times, coast Eskimos had light and heat freely available. For cooking, a soapstone kettle hung suspended above the lamp, near the blubber. Above it was an open latticework of sinew held by a wooden or bone frame something like the web of a crude bearpaw snowshoe. Here the woman of the house dried clothing, virtually a life-and-death responsibility since wet mittens or parkas could bring frostbite and perhaps ultimate freezing for a hunter, or anyone else, whose work kept him outside.

❖   ❖   ❖

Contrast the misery of the early polar explorers with this routine warmth and dry clothing within an Eskimo household. For example, Nansen, sledging across Franz Josef Land after he left the *Fram*, wrote of cold and unwanted dampness: "Saturday, May 30th. Lying weather-bound, stopping up the tent against the driving snow, while the wind flits round us attacking first one side and then another. It was all we could do to keep ourselves tolerably dry, with the snow drifting in through the cracks on all sides, on us and our bag, melting and saturating everything." At the opposite end of the globe and less than a decade later, Sir Charles Wright experienced the same sort of misery during the second Scott expedition, and he expressed it with Nansen's same sense of despair: "If only we could have spared a little oil at night so we could change our clothes. You can shake the ice off, and you have the primus for cooking and the tent to keep away most of the wind; but with a long journey to make you can't afford the weight to carry oil for warming the tent. The primus goes out the minute the meal is done."

Sleeping bags gained up to thirty pounds of ice on that polar journey. "We had to melt the frozen sweat from our clothes at night as we lay in bed," Sir Charles told me, "and down the moisture would go to the coldest part, which is the bottom of the sleeping bag. So when it was cold the bags just filled with ice and that was that. They wouldn't have filled nearly so fast if we'd had separate bags for our heads, to catch exhaled breath. But of course the right thing to do really was not to have gone man-hauling in winter." At best the bags of eiderdown and reindeer skin on Scott's expedition were like slightly warm sponges. In the morning they couldn't be rolled, lest they would be impossible to unroll when night came. They had to be loaded onto the sledges flat, with a wad of clothing stuffed into the opening to give a starting wedge for getting back in at night. The only way to accomplish this was to melt in a few painful inches at a time. "It took an hour to get in," Sir Charles said. "Put your feet in. Melt that much. Shove farther."

Ironically, with all the fighting of unwanted moisture,

getting water for drinking and cooking was so much of a
chore that sledge travelers often complained of "polar
thirst," as do glacier climbers today. "I had prepared myself
for this thirst . . ." Nansen wrote, "and had taken with
me a couple of India rubber flasks, which we filled with
water every morning from the cooker, and by carrying in
the breast we were able to protect them from cold." Eskimos
used a similar system. They kept a pouch made from the
flipper skin of a seal filled with snow and carried inside the
parka where body heat would melt the snow. As a swallow
was taken from the pouch, more snow was added as replacement. Another way was to clean the stomach of newly killed
caribou, fill it with snow, and put it back into the still-warm
carcass to melt. Caribou hair is such a remarkable nonconduc-
tor of heat that a kill covered with snow still is warm the
next morning. Water could also be obtained from snow by
leaving the blubber on a skinned seal and placing the carcass
in a depression scooped into the snow and edged with snow
blocks. By adding a wick to the blubber and lighting it, the
seal would burn slowly, almost like a lamp, and snowmelt
would collect in hollows cut into the snow. Along the coast
another major source of water is sea ice. It loses its saltiness
by the second year.

Understandably, men from the green and blue lands of
central Europe and America, or even from the more nearly
polar Scandinavian and Soviet countries, often find themselves sorely tried by the unremitting rigors of the white
deserts at each end of the earth. Basics of life such as
warmth, shelter, and water—or rather the lack of these—
torment body and mind. Even more threatening to morale,
unless men guard against it, is monotony: Men live and
work in cramped huts buried beneath snow and with no
outside stimulation, not even anything to see "on a lifeless
expanse of barren snow," as the leader of the 1950 Norwe-
gian-British-Swedish expedition expressed it. The darkness
of the winter night, no worse than in Troms or Finnmark
back home, wasn't a problem; neither was the cold. It was
the inescapable sameness that affected morale. Conversely,
among the native peoples of the northern snows, monotony

seems to have been no problem. They had no way to apply the experiences or attitudes of more populous societies to their isolated life, and they had no reason to begrudge the harshness of their land. Even their feelings about the function of a house differed from those of European-American culture. An igloo was not looked upon as a haven for screening off the rest of the environment, except in its role as shelter from the elements; it was never a retreat *from* anything except cold and wind. "Home" was the whole region. A man went indoors for warmth and sociability and sleep, not for privacy or activity, both of which were abundantly available beyond the door.

The traditional dwellings of the North were extensions of the land itself, whether built of snow or of earth, wood, whalebone, or skin. Not so today. The conditions of the arctic environment still rule but even native peoples no longer adapt to them as in the old days. Nor, with exceptions such as the Alaskan North Slope Arco-Exxon-British Petroleum technological island, does man efficiently shield himself. He now neither cooperates nor conquers; he is caught somewhere between. For Eskimos this has been acutely true. Overall they have lost their own way, yet until recently have lacked the affluence needed for the new, outsiders' way. Their own houses, built small, low, and thick-walled, stayed warm, yet not stuffy. Then white men arrived and introduced architecture intended for temperate latitudes. The new style meant flimsy walls too vertical and high to bank with sod or snow as insulation even if custom favored such practice. Convection currents in the old houses brought in air through the open doorway without chilling the whole interior, because the entry always was sunk below floor level, and the cold, heavy air tended to hang there. The new houses, however, lose heat by conduction through their thin walls, and occupants must squander fuel trying to stay warm.

When did man first encounter snow? No sure answer exists, but it may have been as long ago as two to three million years while he was barely beyond the cradle as a species. The place would have been somewhere within the shadow of icy Mount Kenya, Mount Kilimanjaro, or the Ruwenzori Range. Certainly by half a million years ago man had adjusted to the frigid realities of life at the edge of the glaciers that then were sheathing the ice-age Eurasian continent.

Physiologically his survival in a realm of perpetual snow and ice seems unlikely: thin skin, little protective blubber, no fur, spindly limbs. Unclothed and unsheltered he shivers miserably at around freezing, a poor performance compared, for instance, to an arctic fox that can doze snugly in extreme cold, or to a polar bear that simply lets its metabolism drop for the winter and enters the peace of hibernation. Even shrews seem better equipped than man when it comes to dealing physiologically with the realities of the northern winter. According to recent Russian studies, adult shrews lower their weight by as much as 40 percent at the onset of cold, thereby greatly reducing the need for sustenance during the months when snow lies deep and food is hard to get.

Man can make few adjustments to cold within his body, although Eskimos differ somewhat in this from unacclimatized Caucasians or Blacks. Eskimos' hands stay warmer. More than half again as much blood flows to the fingers, which keeps them dextrous even while in slushy snow or water. Tests show that human fingers chilled to a skin temperature of 68° F. are only one-sixth as sensitive as at normal temperatures. This makes the hands next to useless and easily could cost a man his life in the Arctic, but the Eskimos' circulation avoids the problem. Also, Eskimos maintain higher body-core temperatures than Caucasians do during exposure to severe cold, and they recover from the chilling quickly, whereas Caucasians' temperatures not only fall lower at the outset but continue to drop for a while after the exposure to cold is over. Slightly more adipose tissue among Eskimos may help. Probably more significantly, their basal metabolism stays above that of temperate-zone people.

At the opposite end of the globe from the Eskimos, Alacaluf Indians of Tierra del Fuego also demonstrate a high metabolic rate. About double that typical for Caucasians, it lets them work and sleep unclothed and uncovered at freezing temperatures. In the 1820s Charles Darwin, aboard the *Beagle*, wrote of seeing a Fuegian woman nursing a baby while standing in a sleet storm, the white of the slushy flakes melting against both their skins. A few decades later Lucas Bridges, a missionary, told in his book *The Uttermost Part of the Earth* of Yahgan Indian women swimming with toddlers on their backs while mooring their canoes in off-shore kelp beds. The only time one of these small passengers seemed to really interfere with its mother's dog-paddling was in winter when it sometimes tried to climb onto her head to avoid contact with ice-filmed kelp blades.

Walking barefoot in snow was common among the people of Tierra del Fuego and also is reported for Aleuts and Bering-Strait Eskimos. Seemingly, in the Arctic, the practice was to save wear on boots, for coarse or wet snow is hard on footgear, and replacement took hours of work. The Ala-calufs and Yahgans had no footgear, although particularly for the Yahgans, who lived in the southernmost part of the archipelago, snow fell frequently even in summer. Nonetheless, these people's only real concession was to tie on shin guards of stiff guanaco hide when walking through crusted snow. Their feet they left bare, and they wore no clothing beyond a breechcloth or apron augmented by a scrap of sea-otter or fox skin no bigger than a handkerchief. This small garment was tied around the torso with a cord and shifted from front to back and side to side as token protection against wind and driving snow or rain.

Such instances notwithstanding, man's life in cold regions depends largely on contriving ways to spare himself its effects. Physiologically he is doomed. Behaviorally he manages. How comfortable he is, and how favorable his survival odds become, depend in large measure on his success in selecting the right clothing and caring for it properly, as well as in providing himself with shelter. The earliest written record of snow-country clothing seems to be a reference in the annals of the T'ang dynasty of China, A.D. 618 to 906. A note concerning the *Wu-huans* of Bayirku mentions that "the people clothe themselves with reindeer skins." The reference is appropriate as a first recorded observation, for in the North this skin was more commonly used to make clothing than any other. (Reindeer in the Old World and caribou in the New World are the same animal called by different names.) Among Eskimos sealskin also was of great importance, and various Siberian people preferred dogskin. Fox, hare, and eider-duck skins went into clothing, but not for everyday wear; they are too delicate. Polar-bear skin was too heavy for general wear, although hunters stood on pieces of it as they waited in the snow for seals to return to their breathing holes. Musk-ox fur is too shaggy to keep clean.

In areas where sealskin was important for clothing, the animals' intestinal membranes also were used to make waterproof parkas and pants, forerunners of today's rubberized and plasticized outerwear. In other areas the skin of whales' tongues, or the outer membrane of the liver, was favored for this use; so were the bladders of halibut and the intestines of sea lion, bear, and caribou. Fishskin provided still another waterproof material and reports of Siberian journeys in the eighteenth-century mention clothing of pike and turbot skins among the Ostyak people, suggesting that these were their only garments part of the year. "These skins are very airtight and strong," one of the reports claims. "After being rubbed with fat they are perhaps not inferior to skin with fur with regard to retaining heat; and in snow and slight cold they protect even better against moisture than does skin with fur." Skin from salmon and pike served as tent covers among the Tungus, as well as for clothing. Samoyeds used fishskin for caftans and stockings worn as summer garb. Lapps fashioned salmonskin into headgear. Ainus, Kamchadals, western Eskimos, and Athapascan Indians on the Yukon and Kuskokwim rivers used it for outer footgear. This is mentioned as late as 1909 in various reports. So common were fishskin garments that for the Arctic as a whole they ranked second after those of reindeer/caribou hide in importance. They were worn more widely than were garments of sealskin or dogskin.

For boots the skin of reindeer legs was best, and the tough hide of bearded seal was especially prized for soles because of durability. In some areas sealskin was used for uppers as well as soles. It made boots more watertight than was true for those of reindeer hide, even allowing for removal of most of the seal oil in tanning. This waterproof quality offered great value during times of wet snow, such as in fall before the arctic world froze hard and in spring during the thaw. Sealskin boots are said to be as watertight as modern rubber boots—both a blessing and a curse. They stay dry on the outside but sopping on the inside; the dampness of snow can't penetrate and neither can sweat escape.

In an Eskimo or Athapascan village today, even in winter, youngsters sled and throw snowballs while dressed in mail-order jeans and tennis shoes, and adults save skin parkas for special occasions. Both in money and in time, the cost of making clothing the old way has climbed too high. Eighty to one hundred dollars is the price simply for a ruff with double strips of wolf fur back-to-back, stiffened with seal-skin between the two and edged with wolverine fur around the face. Whole wolf skins sell for $150 and up; wolverine skins for around $100. Wolf fur for a ruff comes from across the animal's shoulders. From a wolverine the best strip is up the hind legs and across the back. Each skin gives only one ruff.

Wolverine fur generally is thicker than wolf fur, and it holds up better as frost or snow is brushed off. Wolf hair tends to come out when brushed repeatedly. Also it mats and freezes if damp and thereby loses the air from among its hairs, which diminishes its insulation value. Wolverine will frost up; any fur will as breath condenses on it when temperatures are low. But the texture and oiliness of wolverine fur keep it from matting, frozen or no. Step into the house and the ice falls off by itself from a wolverine ruff, or it can be brushed and shaken off. Not so with wolf fur. It gets increasingly wet as warmth melts the frost or snow clinging to it. The same is true of caribou, and in addition its hair sheds readily. Even as visitors to the Kobuk River of Alaska, my husband and I knew the plague of caribou hairs stuck to cameras, floating in the water bucket, and, somehow, even mixed into the yogurt. Fully haired caribou hide does offer special warmth, however. As is true of all members of the deer family, individual hairs have rather large tubelike centers filled with air cells. This greatly increases the insulation value.

Continual care of native garments accounted as much for their effectiveness as did the design and materials themselves. Clothing simply was not allowed to get damp or ripped. Baffin Islanders had a saying that a woman with children knew no moment's freedom, so busy was she drying and repairing her family's clothing and sleeping robes. Rips and beginning weak places were corrected immediately, and every flake of snow or trace of rime was shaken or beaten out before parkas and pants were put on in the morning, or robes snuggled into for sleep. A special flattened stick was kept handy by the door expressly for the purpose. Dampness was dealt with before it had a chance to happen; snow and frost weren't allowed to melt.

Avoidance of sweating was important, too, and the cut of clothing helped in this regard. A layer of dead air was held between clothing and the wearer's skin and, warmed by the body, it stayed in place. Furthermore, cold outside air couldn't come in so long as a parka was worn closed at the waist (or hips) and throat. If ventilation was wanted, the bottom edge or neck, or both, could be opened to provide a regulated flow of air. This would cool the body and evaporate sweat, extremely important not only for comfort but also to keep the clothing from getting damp. Without pulling his second parka on or off, an Eskimo hunter could be warm enough to ride in a sled, or cool enough to cut blocks for a snow house or to paddle a kayak against the current without sweating. Eskimo women's parkas were cut big enough across the shoulders for a baby to nestle naked against its mother's back and be swung around to nurse without leaving the warm microclimate of the clothing. Only when a baby started to urinate did it get removed, and then with astonishing haste no matter what the temperature happened to be.

Clothing serves three purposes in the white world of the polar and subpolar regions and the high mountains: It insulates the body, keeps it dry, and protects it from wind. The whole idea is to hold in body heat and, whether with an Eskimo's caribou parka or a mountaineer's down jacket, this is accomplished by compartmentalizing air into small, separate units within the fur, fiber, or fabric of each garment and between layers of clothing. Getting wet from the melting of snow or from sweat can be disastrous in part because the moisture fills the dead-air spaces and destroys the insulation. Wind spells trouble because to be effective dead air must stay "dead"—stationary.

Among fabrics, wool for centuries has been the warmest in existence largely because of dead-air traps naturally present in the fiber. A scaly outer layer encloses the cortex and is itself surrounded by an outer cuticle. Minute air spaces lie between the scales, walled top and bottom by cortex and cuticle. Sheep's wool is the most widely used in contemporary western society, and one of winter's traditional odors has been that of wet wool drying when family members come in from playing in the snow or from shoveling the walkway or carrying feed to the livestock. Today synthetics compete with wool but according to Jim Whittaker, who manages Recreational Equipment Incorporated in Seattle, good wool knickers and shirts are returning—and they're not just for mountaineering or skiing use. "Students are wearing survival gear on campus. It's good judgment."

For Jim, hailed as the first American to climb to the top of Mount Everest, pairing the words "survival" and "good judgment" comes naturally. He was a mountain guide at Rainier during the time we lived there, and on its snowfields and glaciers he honed the climbing skills that have taken him to the highest slopes on earth. At Rainier, too, he first learned the value of preparedness which, as he puts it now, can keep an emergency from being anything worse than an inconvenience. Recently I sought him out to ask about current trends in snow clothing. REI equipment goes into snow all over the world now; the corporation, a cooperative, is the biggest outfitter of individual outdoorsmen and mountaineering expeditions anywhere, and, in Jim's words, "Going into mountains means going into snow. You learn to like it, or at least how to handle it.

"What you need depends on whether you'll be in dry snow or wet snow. I'm always glad when I get far enough up a mountain to start hearing the snow squeak underfoot. That means it's cold enough to be dry. Down jackets and pants are the thing to wear then—except while cooking in a tent. That gets everything steamed up, but usually you can hang clothes and sleeping bags outside to freeze-dry; just shake out the crystals a few minutes later. Wet snow is more of a problem. When the temperature is near freezing and snow is wet, down is likely to get damp and collapse. It's useless. What you need then is something that will stay fluffed out even if wet, and that means wool or Fiberfill II."

Proud of REI's quality-control standards, Jim himself recently had tested a Fiberfill II jacket by wearing it into his swimming pool until it was soaked, then walking a nearby beach in it for four hours. Direct from the swimming pool, "That jacket must have weighed forty pounds, but I stamped out most of the water." Stamping is better than wringing because it doesn't damage the delicate fibers, which are hollow filaments of Dacron. Since they are noncellular they don't absorb water as wool or down does and consequently Fiberfill II, or the similar Polar Guard, offers some warmth while wet and dries quickly. With the soggy jacket Jim tested, he wore only blue jeans, yet was comfortable despite a cold winter day.

Protection for the feet probably stirs argument more quickly than any other part of snow clothing. Separated farther from torso warmth than any other part of the body, the feet also are the most likely to be in contact with moisture if snow is wet, and they are great sweat producers on their own. "Anything rubber is all wrong," a Yukon trapper once told me. "How are you going to dry it?" And when he saw a pair of eight-pound U.S. Army boots developed for use in Korea, his astonished remark was: "Those must be terrible. You couldn't go anywhere. All you could do is sit in the truck and steer." For running a winter trapline he relied on high-top moccasin boots worn with a felt liner and four pairs of wool socks; yet Jim Whittaker listed Korean boots as a standard for general use.

Preventing a bridge of moisture between foot and the outside air or snow surface is important for staying warm, and Korean boots, with their double layers of insulating felt and vaporproof rubber, achieve this. A double layer of plastic can provide the same effect, although not completely comfortably so. Climbing at Rainier we sometimes put grocery-store plastic bags over our bare feet, added two or three pairs of wool socks, and then a second bag over them. The first layer of plastic held in sweat; the second kept out external moisture. The socks in the middle cushioned and insulated. Our feet got wet but stayed warm. For day after day the system wouldn't be wise because the feet might suffer from such constant tropical clamminess. But for short periods plastic bags offer a no-cost guarantee of warm—and wet—feet.

For tents and weatherproof outerwear, a new fabric called Gore-Tex prevents the penetration of outside moisture, yet permits escape of "the damp exhalations of the body," as Nansen termed sweat and breath. It is a laminate of three layers: ordinary uncoated nylon on the outside, a middle film of Teflon, and an inner layer of knit or woven nylon or of polyester. The Teflon has microscopic pores that vapor particles can get out through but that wet snowflakes or raindrops can't soak into even when storm-driven. Previous nonporous material has held in sweat and breath and caused problems that range from appalling misery to the merely less-than-ideal performance of our Rainier plastic socks.

Even the best in clothing may not be enough. Jim mentioned being caught without oxygen on Mount Everest at the 27,500-foot level and feeling his inner fire start to go out. "Clothing can hold in heat the body produces but if the furnace within is going out, no jacket or sleeping bag can help. And to keep the furnace going you need fuel and oxygen. You have to eat and to breathe. If you haven't food, or oxygen, blood leaves your extremities and moves to the body core, and you get frostbite and maybe die. The bigger your body [and Jim is six feet, four inches] the more food and oxygen you need."

Given them, the principle involved in dressing for snow and cold is the same whether for an Everest climber caught by nightfall, or avalanche, or for an Eskimo hunter waiting by a seal's breathing hole or a city dweller who has forsaken his apartment in favor of a weekend on the ski hill: Provide insulation and control moisture.

CHAPTER

6

BLIZZARDS AND AVALANCHES

RECORDS OF SNOW DISASTER DATE WELL BACK INTO HISTORY. Hannibal crossing the Alps with his elephants in 218 B.C. still comes readily to mind because of the avalanches he experienced. Full details aren't known (not even which pass he used), but much of Hannibal's story has come down to us. His purpose was military, and the crossing took place "at the setting of the Pleiades," probably late October. His entourage included at the outset thirty-eight thousand men, eight thousand horses, and thirty-seven elephants, but before the crossing was completed nearly half of the men, a quarter of the horses, and "several" elephants had fallen victim to mountain snows. For two days Hannibal's expedition camped at the top of the pass; then they started on. A blanket of fresh snow covered the crusted snow of an earlier storm—notorious avalanche conditions—and, as the descent began, the animals' feet perforated the upper fluffy snow. That layer gave away and men, horses, and elephants plunged helplessly downslope. The tragedy has become one of the world's classics.

I thought about it a few years ago while crossing Brenner Pass in a blizzard. We had spent the night at a farm pension outside Innsbruck. Rain against the window at first acted as lullaby, then its pelting changed to the silken whisper of snow, and by dawn we knew we should hurry. Other travelers also were defying the storm, and a certain camaraderie prevailed as road conditions sent one after another of us into a skid. We all would stop to help, holding counsel in a babble of languages and then somehow uniting to shove each car from ditch back onto road. Slowly misfortune—or wisdom—thinned the line of cars climbing the pass, and the quietly deepening snow blanketed a realm progressively emptied of travelers. We continued in our Volkswagen, however, until perhaps five miles below the summit. Then we too slid off the road.

Practicality outweighed chivalry in importance so I got out to push, leaving my husband, a better driver, at the wheel. Since the car was light we easily got going again, but as the wheels took hold a sudden realization came to me. There could be no stopping to let me back in, lest we again slip off the road. The only thing for me to do was leap onto the back bumper, sprawl against the slope of the car, and brush snow from the rear window to let Louis know I was aboard. In this fashion we pulled up at Italian Customs —and if officials there were surprised to see a snow-plastered woman in high-heeled shoes disembark from the rear bumper they gave no sign. My own thoughts, aside from self-pride in quick thinking, had been of Hannibal. I wondered whether this was the pass he had tried, and I empathized with his disregard of the odds. He, as we, had doggedly continued on a previously planned course, which is behavior more common than admirable.

The human refusal to surrender to reality runs a stubborn thread throughout the literature of snow, for most persons caught in the white disaster of blizzards or avalanches somehow deliberately court their fates. The turn-of-the-century Klondike stampeders furnish an example. As someone has written, God must have chuckled when he created gold: Nothing, not even women or drink, turns men into greater fools. Certainly in thirsting for wealth most of those involved in the gold rush of 1897–98 disregarded topography, climate, and their own human frailty.

Announcement of the newly discovered riches of the Klondike prompted a rash of advice on how to get there, and chartered boats began unloading would-be miners at a variety of unpromising access points. Best known is the route over Chilkoot Pass which led thirty-two miles from the coast at Dyea, near the top of the Alaskan panhandle, to Lindeman at the headwaters of the Yukon River. Seventy feet of snow fell at the pass during the first winter of the stampede; one single night brought an awesome six-foot snowfall. Tents collapsed beneath its weight, smothering occupants and burying caches of goods stacked layer on layer and futilely marked with poles by men relaying loads up the long steep slope to the pass. While the heaviest storms raged, men of the Royal Canadian Mounted Police who were stationed at Chilkoot shoveled snow around the clock to keep their own roofs from collapsing. Even during mild periods the pass, which straddled the Alaska–British Columbia border, was an unpleasant place to be. Firewood sold for a dollar a pound, and a cup of coffee with a doughnut cost two and a half dollars. Mounties required each man entering Canada to have with him a half ton of food, an amount judged sufficient for one year, and they charged duty on it all; sixty cents for each barrel of flour, oatmeal, or cornmeal and varying rates for boxes of matches or sacks of pitted plums or tins of fish.

Getting to the gold fields was physically and financially exhausting, yet two thousand men including Canadians, Yankees, Greeks, Scotsmen, Japanese, and Kanakas (Hawaiians) checked through customs the winter of 1897–98. Women came too, including one seventy-year-old German matriarch who wore an ankle-length dress covered with a

lace apron. There also was an incredibly sturdy and enter-
prising female who traveled alone, tugging a sled loaded
with a small glowing stove. From time to time this woman
stopped to warm herself and make tea, or a meal, of course
offering to share the repast with her less imaginative male
companions. Their minds were fixed on washing raw gold
from the gravel; hers settled for mining the wealth of pants
pockets.

An avalanche struck on Good Friday and buried hundreds
of those climbing Chilkoot Pass; yet it killed "only" sixty-
three of them. An ox buried by the snow was dug out alive
and used to haul bodies to Slide Cemetery at Dyea, where
they still lie. Despite all this, however, the most dreaded
hazard wasn't the snow but the Canadian customs post at
the top of the pass. To avoid it non-Canadians often let
themselves get suckered into alternative routes that avoided
British Columbia soil and officialdom and led directly from
the coast to the Klondike, which some believed to be in
Alaska and others knew lay within the Yukon just beyond
the Alaska–British Columbia border. Two of the routes alternate to the Chilkoot led over glaciers: the Valdez and the
Malaspina. Twice the number of men who braved Chilkoot
Pass that first winter after the strike set out over the Valdez
Glacier, even though as an approach it led across a giant
tongue of crevassed ice that stretched for twenty miles
from Prince William Sound lowlands to a forty-eight-hun-
dred-foot mountain crest, then down for nine miles to a
series of streams and lakes that fed ultimately into the
Copper River. The river, in turn, could be followed to the
gold fields.

Onto the snout of the Valdez ice the men wandered, gold-
crazed, inexperienced, uninformed, and ill clad. To haul a
ton or so of gear apiece they brought burros, mules, horses,
dogs, and sleds. At least they started with them, but rough
water en route to the glacier often forced the shooting of
scores of animals, which were thrown overboard along with
all hope of using their strength to reach the Klondike. So few
animals survived that horses purchased for fifteen dollars
in Montana sold for five hundred dollars in Valdez, and

sleds of every design and construction also commanded exorbitant prices. One group of eager prospectors imported an untested steam sled, which they spent weeks assembling only to find it wouldn't draw its own weight on the level, let alone pull a load up the glacier.

Even the most stalwart men must have felt daunted by the reality of where their enthusiasm had led them, but turning back was not readily among the options. On they pressed, drawn by the rainbow beyond the ice. Snow six feet deep greeted the first arrivals; those who came later in the season found more. Travel by day brought the torment of snowblindness, which seemed to fill the eyes with red-hot sand and turn the white world to quivering crimson. Travel by night was better, if colder. The snow crusted enough to bear a man's weight by nine o'clock, and by midnight a horse could walk without sinking in. Inadequate diet brought on scurvy and, since much of what food there was had half rotted, nausea was common. Appalling sanitation prevailed. Exhaustion ran rampant. Bitter quarrels sprang to instant life among partners whose chief bond never had been more than shared avarice. Stupidity ensued. A trio with two pairs of jointly owned oars are recorded as taking one apiece and destroying the fourth. Three other partners broke a grindstone, and each stomped off with his useless third.

Downslope winds coated men struggling up the glacier with wet snow, and blizzards drove flakes into whatever shelters could be improvised. By spring, rains set off avalanches. When summer came crevasses in the ice opened and travel—mercifully—became impossible for all who hadn't yet started up the glacier. All who had succeeded in relaying equipment to the top, however, now were faced with relaying it down, and disaster loomed even more on the descent than on the ascent. Sleds plunged out of control and in trying to stop them, men got run over. Best estimates place success via the Valdez Glacier at about one half of 1 percent of those who tried it. The government issued rations to hundreds of men who lost everything; the Christian Endeavor Society built a relief station to rescue those too worn out to continue farther across the ice or to cope with their defeat; and a whaling company carried home scores of the forlorn.

Fortunately, not as many Klondikers attempted the Malaspina Glacier as the Valdez. At fifteen hundred square miles it is the largest piedmont glacier on the continent and an incredibly rough route. Nonetheless, perhaps one hundred men landed at the Malaspina in the spring of 1898, forty-two of them destined to die there and others to be crippled for life. One party of eighteen recruits who had answered a *New York Herald* advertisement spent three months crawling over the ice in company with the man who had placed the ad. They slept in the lee of their equipment because nights were too windy for tents, and all were so weary they didn't speak for days at a stretch. "It was like traveling with deaf-mutes," noted the leader, Arthur Arnold Dietz. One sled, loaded with the party's most valuable possessions, disappeared into a crevasse along with four dogs and Dietz's brother-in-law, who was driving. A second sled and dog team fell behind in a blizzard, then vanished after the driver had gone insane from snowblindness and frustration. With the provisions of these two sleds gone, dried beans remained the only food, and a third man soon died of malnutrition accompanied by agonizing indigestion. Ice cut the feet of all, yet all not only continued but even dragged a quarter-ton motor up the glacier only to abandon it on the other side when winter's darkness and snow settled in. Crazy with the boredom of a bivouac, four men struck out for Dawson and were lost. The other twelve huddled into a log shanty and spent the brief, gloomy subarctic days and the long nights cutting wood and stoking the fire. When spring came they started back over the mountains, hoping for escape.

An avalanche took three, scurvy a fourth and fifth. This left seven who somehow made their way back to the beach, where men from the revenue cutter *Wolcott* found them. By then three of the seven lay dead in their blankets and of the four who lived, two had totally lost their eyesight from snowblindness, and the other two could barely see. All were dazed and uncomprehending. Despite this, a *Seattle Times* report of the rescue mentioned the men's "success," for some-how they had managed to acquire and keep a little gold dust.

To South American gauchos the word for blizzard is *el tormento*, expressive of the winter anguish experienced around the world each year. This havoc is dramatic enough that a Manhattan office worker aboard his morning commuter train or vacationing in Florida may read, for example, of cold air from Siberia howling into Europe and from his own experience empathize with the unfortunate whom winter's caprice has singled out this time. One such outpouring of cold in February 1956 halted more than one hundred Brussels taxis and forced Amsterdam milkmen to stop deliveries because milk frozen in the bottles kept breaking the glass faster than they could make their way through the snow-clogged streets. Across the channel Royal Air Force helicopters rescued swans held fast by ice in Suffolk and crofters in Scotland, hopelessly cut off by snow, received merciful airdrops of food. A six-inch blanket of white mantled Rome, and even Jerusalem lay covered by snow. During the opening weeks of 1977, notorious in several places for record snowfall, similar situations occurred, varying from inconvenience to disaster.

Suffering can be dreadful during violent snowstorms, yet Stefansson, writing in *My Life with the Eskimo*, says that a blizzard need not be dangerous even for one stranded out in it. The main thing is not to fight the storm foolishly; if visibility is nil, wait for a bit of clearing. He gives four rules to follow during the wait: Keep still, moving only enough to stay warm. Take care not to overexert and sweat, which dampens clothing. Contrive a shelter of some sort, preferably of snow, if the temperature drops to −10° F. or lower. Sleep as much as possible.

This last counters all usual advice, but Stefansson insists it is the Eskimo way. These natives of the snow stop as soon as they realize they are lost, deliberately doing so while still physically and psychologically strong. If possible they build a snow house; if not they sit facing out of the wind, head resting on knees. Either way they doze off to pass the time, and when they feel chilled they waken and exercise just enough to warm back up. They avoid real exertion which might lead to exhaustion, or to wetting the clothes with sweat, a potentially fatal error. Stefansson tells of an Eskimo woman he knew who was caught by a blizzard when no more than half an hour's walk from home, yet stopped and waited out the storm for three days and nights. When the snow finally stopped blowing she walked on and entered her igloo unharmed. She was without shelter the whole time but reported having sat quietly, sleeping when possible and walking slowly about whenever she wakened and felt cold.

As close to home as she was, her realization of the danger in trying to continue was wise. Blizzards can blot out even the closest and most familiar of landmarks, and snow may fall so thick it is hard even to breathe without sucking in its flakes. If the temperature is low under these conditions the snow usually is fine and dry and sifts into any least opening. Frank Debenham, a geologist on Scott's second, tragic expedition, tells of watching a pinhole in the tent wall during a blizzard and seeing "a small trickle of snow dust coming through it looking exactly like the trickle of fine sand running through an hour glass." Similarly, a settler near Klamath Falls, Oregon, in the late 1800s speaks of snow passing through double storm windows even though they were caulked with cotton, and she had to plug the keyhole of the door "or the snow would blow through so bad during the night that by morning it would be drifted entirely across the room, from keyhole height at the door to floor level at the opposite side of the room."

With a wet blizzard the problem usually is sticking rather than drifting. Crushing snow loads build onto roofs and "shoveling" means using a saw to cut it into icy blocks than can be toppled from the eaves. Each winter at Mount Rainier we skied into the back country to clear snow this way from patrol cabins designed with roofs too flat for the snow to slide by itself, and while we spent the first night in each we would fall asleep to the creaking complaint of the overloaded timbers above our beds. When roofs are pitched properly they periodically unload on their own, and the snow slides with a roar that from inside the house sounds frighteningly like the roof has collapsed. Snow-country residents know not to stand or park a car close to a building with a loaded roof; getting caught by snow sliding off is akin to being in the path of a dump truck when it unloads wet concrete.

Perhaps ranchers have the hardest time of all with blizzards, for they must try to care for livestock as well as for themselves. Wet snow encases range animals in white, plastering shut even nostrils and eyes. The daughter of a pioneering North Dakota rancher we knew in the Badlands includes among her childhood memories winters when "the men would rope themselves together and follow the fence to get to the shed. There they wiped the stuck-on snow from the faces of the cattle so they could breathe." Even after such a blizzard had died and the sun was shining, "some of the cattle would have to have their eyes wiped out so they could see to come to the feed that was thrown out for them." Inevitably, others would be found dead. Losses within a county at times ran into the tens of thousands of animals and men would be busy for weeks "after the snow had melted riding over the countryside and skinning all the many animals that had suffocated and died from the snow. Some would be found standing up buried by a drift with nothing but the horns visible."

Given shelter, the animals usually fare all right: "The first thing to do when the blizzard let up was to get into the barn and milk the cow. Then the milk had to be taken to the chickenhouse where the calf was being weaned. The men couldn't see the chickenhouse, it was so completely covered. But they shoveled where they thought the door was and dug a tunnel. When they opened the door, steam came out so thick it was like a fog. The snow had kept the chickenhouse that warm! The little calf was as wet as if she had just had a bath."

Stefansson, too, spoke of North Dakota snow. He was born on a Dakota farm in 1879 and at about age eighteen had started a ranch with three other boys. Their first Thanksgiving a storm from Canada turned the world to swirling white. Somehow the boys had not strung a rope from house to barn, as was the custom, but Stefansson decided to brave the blizzard and care for the animals anyway. He thought he could use the wind direction as compass and, since the barn was a long one standing broadside to the house, he was confident he could find it. He did but then couldn't find the door. It was on the lee side of the building, and a drift there hid all traces. Shoveling would be pointless; snow would drift back as fast as any amount of digging could clear it. Besides, the shovel was buried. Stefansson thought of breaking in through the roof but realized he still couldn't carry hay to the animals. The haystack was totally inaccessible. Chastened, he reversed course and arrived safely back at the house.

Years later, writing about the Arctic, Stefansson referred to this event which he had given little thought at the time. "I now consider it one of the most foolhardy enterprises of a career that has been in considerable part devoted to similar things," he wrote. During that storm more than twenty neighboring households experienced tragedy. Men had "gone in search of their barns, never found them, and been frozen to death. Others found their barns and stayed there until the gale was over, not daring to return to the house. Still others found their barns, fed their stock, and lost their lives on the way back to the house. There also were stories of lightly built shanties that had been blown away by the wind, exposing the occupants to the blizzard or killing them in the wreck. At that time I agreed with all our neighbors . . . that gales such as I have described were exceedingly dangerous to life and limb. That was because we did not know how to deal with them. I have since learned from the Eskimos how to get along in a blizzard and I should feel ashamed of myself if I suffered anything as serious as a frostbite from a day out in one." Defying the blizzard and going to the barn wasn't what Stefansson had come to see as foolhardy; it was his unknowledgeable handling of the effort.

Wild animals have certain safeguards for the snow they are sure to encounter. Polar bears' eyes, for instance, are equipped with a nictitating membrane that passes sideways across the eyeball and clears slush, which blinds prairie cattle and arctic sled dogs. At the onset of winter sables outside Moscow grow forty-four fine underhairs for each longer bristle hair, twice the number of underhairs they have in summer and marvelous protection against cold and wind-driven snow. Martens achieve the same effect in a different way. The underhairs of their fur nearly double in length, and the new bristle hairs grow in one-third longer than those of the summer coat. This means more air trapped within the double winter coat and, therefore, increased insulation. Arctic foxes turn white in winter, as camouflage, and the shafts of the new hairs hold minute dead-air spaces for insulation. The soles of their feet become furry, which simultaneously helps keep the paws warm and acts as built-in "snowshoes."

Against one aspect of winter, however, there is little protection for animals that chance to be exposed. This is the white fury of an avalanche. Popular belief holds that mountain-dwelling animals such as old-world turs and new-world mountain goats (which actually are one type of antelope, not a true goat) instinctively save themselves from avalanches. This isn't true. They often recognize the deep roar of onrushing snow for what it is, but they can't always save themselves. Some mammalogists believe that avalanches take a greater toll of these animals than any other one cause of death. Nearly each summer at Mount Rainier one or two mountain goat carcasses melted out of the avalanche debris at the base of one particular slope and in parts of the Alps the carcasses of turs and chamois crushed by avalanches provide a routine food source for vultures and jackdaws.

Avalanches affect wide areas during winters of heavy snow and in such years mountain-animal deaths increase. In the Alps chamois graze slopes where avalanches recently have run, exposing dry vegetation, and while feeding they get hit by new avalanches. As many as thirty carcasses have been found in the snow of a single slide. In the Bzybe Range of the Caucasus, a series of avalanches one winter blocked a river with debris 350 feet deep, and it took years to melt completely and release all of the carcasses of deer and chamois that had been caught. Often such areas of snow-sealed carrion show the prints of lynx and bear that have come to feed.

Aboriginal men probably seldom were affected by avalanches. They had little reason to venture into high mountains, especially in winter. Seemingly mountain men were the first in North America to experience avalanches often, but their difficulties, as those of their Indian predecessors and contemporaries, are little chronicled. Missed at a fur-trade rendezvous, a victim of the snow simply was remarked upon as having "gone under" and that was that. It wasn't until mining towns—actual settlements—boomed in the Sierra and the Rockies, the Klondike and the Cariboo, that Americans and Canadians paid much attention to snow hazards.

In Europe the situation was different. The Alps, five hundred miles long by one hundred wide, have for centuries been the world's most inhabited mountains and they also have the longest written record of mountain conditions. In his *Geography* the Greek-raised Strabo, writing a century and a half after Hannibal's battle with mountain snow, speaks of danger in the Alps "which comes to all, including the beasts of burden, who travel the passes on foot." He concludes that "these places are beyond remedy, so are the layers of ice that slide down from above—enormous layers capable of intercepting a whole caravan and of thrusting them altogether in the chasms that yawn below." Regardless, travelers continued to cross the Alps.

By the middle ages many of them were on pilgrimages to Rome. In December 1128 the abbot Rudolf from Saint-Trond, which is near Liège, led such a band over Saint Gotthard Pass. "As though fixed in the jaws of death we remained in peril by night and by day," he wrote. Avalanches

had them terrified to proceed and equally terrified of lingering. Guides in the little village where they had stopped
refused to risk going on, then finally were bribed into it.
Wrapping felt cloths over their faces and pulling on heavy
jackets, mittens, and spiked boots, they set out against their
own judgment to scout the route while the abbot led a mass
for their safety. Soon, Rudolf's history continues, "a most
sorrowful lament sounded through the village, for, as the
guides were advancing out of the village in one another's
steps, an enormous mass of snow like a mountain slipped
from the rocks and carried them away. . . . Those who had
been aware of the mysterious disaster had made a hasty and
furious dash to the murderous spot and having dug out the
men were carrying back some of them quite lifeless, and
others half dead upon poles, and dragging others with
broken limbs."

Such events must have been commonplace and well
known. Yet both travel over the passes of the Alps, and
settlement in the high valleys, showed no letup. A Latin
account of 1450 tells about a hamlet in the French Alps
where all were killed "save for one family which continues
to live there with great difficulty." Men are said to have
muffled the bells hung around their mules' necks to silence
the clanging lest noise set off an avalanche, and village
children learned not to shout or slam doors on days when
the snows hung precariously to the slopes. In Bavaria villagers each spring painted the sign of the cross onto eggs
and devoutly buried them at the foot of known avalanche
paths, praying divine protection; and a 1652 court of law
officially pronounced witches as the cause of avalanches and
set death as the punishment.

Human action often actually did contribute to avalanching. Villagers in need of lumber and firewood cut forests
on slopes conveniently at hand and thereby created avalanche infernos where the hazard would otherwise have
been moderate. Families then grazed cattle or sheep on the
cutover slopes which stopped all possible forest regeneration
and destroyed bushy growth that might have helped to
anchor winter's buildup. Disaster commonly resulted. There

is a record of two villages that were devastated at the same time by avalanches brought on by cutting the forest, then squabbled bitterly over which one should have the additional timber brought down by the snow.

Today in parts of the Alps man's effect is even more critical. Where ski resorts have taken over farm villages, livestock no longer is led to pasture on the high slopes, and the removal has influenced avalanche conditions in two ways. Paths formerly cut into the earth by the animals' hooves gently terraced the slopes every foot or two, and this was corrugation enough to help hold the snow. Also stock ate the grass to stubble, which left a serrated mat to bond well with snow. But in recent years grass has been growing tall in summer, drying in autumn, then bending to the ground beneath the weight of winter's first snow. This produces a slippery surface that contributes to avalanching. To prevent this some resorts and many peasant villages strictly enforce the mowing of autumn grass regardless of grazing. Forest practices also are closely and specifically linked with avalanche prevention programs, since snow is much less likely to break loose on a forested slope than on a bare one, and trees lessen the force of any slides that do slam down into them.

The siting of villages in the Alps long has avoided known avalanche paths, and as added precaution all residents join in maintaining barriers built on the most troublesome slopes. These are a series of platforms angled out from mountain-sides where there are no trees. Formerly the barriers were built with stones brought down by previous avalanches or from wood, but now they usually are of aluminum which is lightweight, sturdy, and weatherproof.

Such work is done cooperatively; village life—in fact survival itself—has little place for individualism. Communal action dominates, several aspects of it quite directly linked to avalanche danger. Men get together to stamp paths into snow threatening village cows, then move them to safer elevations. Land ownership is so fragmented that a typical holding seldom totals more then ten acres which may be divided into fifty separate, widely scattered parcels. In case

Wind affects snow more than any other external force, except
for temperature. It packs falling snow into tree branches and swirls
fluffy snow into saucers, as eddies first spiral particles upward, then
back into the main windstream. Wind saucers can be hazardous for
skiers and snowshoers who venture too close; their bottoms are soft
and sides straight. Climbing out is difficult. *(Ruth Kirk)*

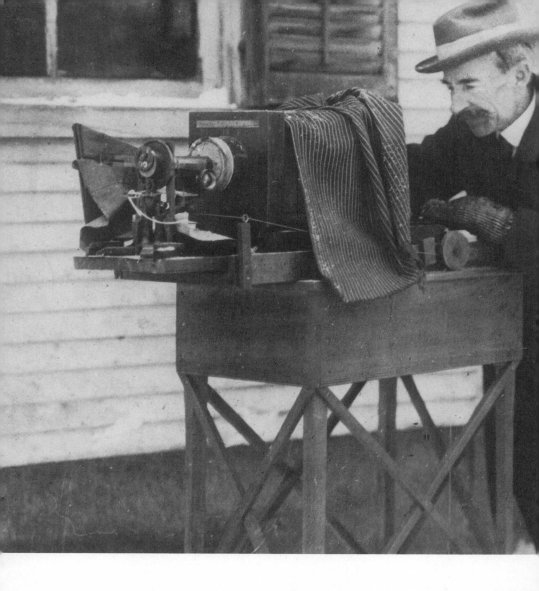

Using glass-plate negatives, Wilson A. Bentley worked outside
during Vermont storms to photograph snow crystals.
How each is shaped depends mostly on temperature and the
availability of water vapor as it forms. In 1931 a book of
Bentley's with 3,000 photographs gave the public its first look
at the infinite variety of snow crystal patterns *(right)*.
(National Archives)

Approximately thirty years of fairly uniform snow accumulation
shows in cross-section at the Blue Glacier on Washington's
Olympic Peninsula, each layer marking one year's snowfall *(left)*.
Dust accumulated on the glacier's surface and concentrated
by summer melt helps to delineate the layers.

In the 1960s University of Washington and Cal Tech scientists
cut a vertical shaft fifty feet into the Blue Glacier *(right)* and erected
a framework from which they could descend by bucket into the ice
to cut samples for density and tensile strength studies. The fifty-foot
depth allowed study of only five years' of accumulation. *(Ruth Kirk)*

Three-foot pinnacles called sun cups form on the upper slopes of 14,411-foot Mount Rainier, Washington. They form because of uneven heating and melting, usually as dry air moves across a snowfield with irregularities already existing. As hollows develop, ongoing melt concentrates dust particles in them. That darkening increases heat absorption and deepens the cups still more. *(Ruth Kirk)*

The United States record for deep snow belongs to Paradise,
a mile-high shoulder of Mount Rainier, Washington, which faces
into the prevailing wind. More than 1,000 inches of snow may fall
in a single winter. Workmen commonly tunnel through snow
in July to open Paradise Inn for the summer season. *(Ruth Kirk)*

Wheeled vehicles cannot travel readily over snow, a contrast
with earlier reliance on sled runners. Snowblowers clear roads
by shooting snow out over embankments. *(Ruth Kirk)*

A record blizzard swept across the northern plains in 1887,
killing cattle, bankrupting ranchers, and greatly
disheartening settlers. The first railroad rotary plows
were tested that same year. *(National Archives)*

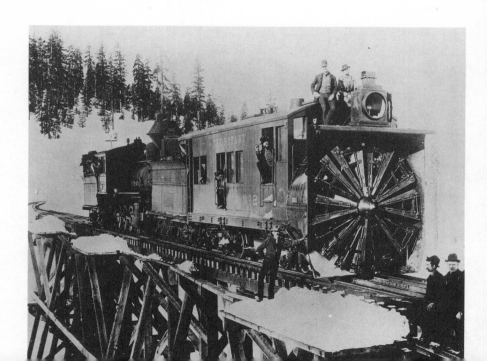

of avalanches this has advantages. A farmer might find one or two of his plots covered by snow or rocks and rendered worthless until he can clear them. But he never will be cut off from his farm altogether. Several plots are sure to remain unharmed, and he can work at a steady pace through the whole growing season. Only winter brings respite to the people of the Alps. The young use the time to court, restricting themselves mostly to their own village, since avalanche danger often makes travel folly. Marriage within the home village becomes the rule and inbreeding is common. It, in turn, fosters additional village solidarity, which leads to still more in-village marriage.

Snow roaring downslope engenders enough force to wreak great and often capricious havoc. A German folk saying asks: What flies without wings, claws without hands, and sees without eyes? Answer, the avalanche-beast. The riddle is apt, for avalanches are capable of stupendous yet selective devastation. A 1951 avalanche in a Swiss village, for example, hit a pile of debris from a previous avalanche and, riding a cushion of air created by the impact, split into three monstrous tongues. One demolished the pastor's house, another slammed into a cowshed and crushed nine cows and six sheep, and the third flattened a villager's house scores of yards distant from the damage of the first two tongues. No other buildings were harmed; no people were hurt.

Often an avalanche will sweep down a slope, across a valley floor, and up the opposite slope, occasionally even ricocheting back and forth three or four times, laying waste with each awesome rush. Pressure as great as eleven tons per square foot has been recorded, a force exceeding by five or six times the battering of storm-driven ocean waves. This high reading for an avalanche was taken with some difficulty a few years ago near Zuoz, Switzerland. Gauges had been placed in a chute notorious for avalanches, but repeatedly the actual pressure of the unleashed snow so far exceeded anticipation that the gauges were useless. An avalanche measured elsewhere at just under a quarter ton of pressure per square foot had lifted and tumbled a railroad locomotive and cars. The power of the Zuoz avalanche, when finally measured, exceeded this force by a factor of forty-four. Nothing in nature outdoes new-fallen snow as a symbol of peace and purity. And nothing holds quite the same Jekyll-and-Hyde capacity to turn serenity into chaos.

Snow can flow almost as a liquid, or stretch elastically, or compress—all without losing its structure as a solid. It can cling particle to particle, layer to layer, and snowbank to ground surface; it also can abruptly lose all semblance of such cohesion. As much as one million cubic yards of snow have been known to give way and thunder downslope at a time. This amount would fill the beds of 10,000 ten-yard dump trucks, so many that if they were lined up bumper to bumper they would reach for about 200 miles. Such a slide may be triggered by nothing more than a falling stone, the hopping of a rabbit, thunder, the sonic boom of a jet airplane, or a man slicing across a slope on skis. Results sometimes are instantaneous, other times delayed. Several skiers may safely cross a poised slope before the cumulative effect of their passing sets off the snow and catches the last members of the party. Or a rupture may streak through compacted snow from a point of first impact to break the anchorage of nearby snow and start a second avalanche. Fractures can propagate as fast as 350 feet per second. In some areas interconnected gullies trigger one another as sliding in one suddenly withdraws support for the snow in another.

The middle elevations of the mountains tend to be the hardest hit. Comparatively little snow falls at lower elevations, and also those slopes usually are forested and for that reason alone somewhat protected from avalanching. Higher elevations tend to be barren and faceted with crags and steep faces that unload themselves before snow builds deep enough for major avalanching. Also, in these highest reaches, winds blow so ferociously that they sweep snow away rather than pack it into great slabs and cornices. Consequently the middle elevations with thirty- to forty-degree slopes generally produce the most avalanches. Convex slopes are the most deadly of all; snow on their steep lower portions tends to pull at the rest of the cover and set up enormous tensile stress. Steepness isn't requisite to sliding. If enough melt-water or rain lubricates individual grains, wet snow may avalanche no matter how gentle the gradient. A mere six or seven degrees can be enough. Or, sometimes instead of lubricating individual grains, the water seeps between layers and causes them to slip. On a moderate slope avalanche danger is likely to be great whenever the liquid content of the snow reaches 10 percent or more. If a handful of snow packs easily in the hand and feels slippery, it probably had reached this point.

Two types of friction play roles in determining when a slope will let go: static friction and kinetic friction. At certain angles static friction holds snow grains together. Other, greater angles are needed to hold them despite kinetic friction or to keep them moving once a slide has begun. Effects vary depending on type of grain. Those that are rounded slide at the lowest angles of friction; those with sharp edges slide only when the angles are larger. Delicately branched snow stars usually interlock when they first fall, forming a feltlike mat that can cling by static friction to a nearly vertical surface; but the rays soon disintegrate, slipping begins, and once in motion such crystals shatter and become fragments even if the angle of kinetic friction is as little as seventeen degrees.

Wind packs snow as soon as it falls, rounding crystals by tumbling them against one another and producing a packed, fairly dense layer. Without wind, snow crystals may be no more than 2 percent of the mass of a snow layer. The rest will be air. Such a fluffy blanket has only about one-fifty-thousandth the hardness of wind-packed snow, although this varies with the force of the wind and the nature of the snow. Different crystal shapes tend to fit together differently. Needles and granules pack more closely than hexagonal plates and columns. Stars lie the loosest of all. Rime also affects the bonding, especially in maritime climates where snow crystals usually reach earth well coated with rime. They act as nuclei for vapor droplets to freeze onto as they fall through supercooled clouds. Such rime-coated crystals bond closely and result in relatively cohesive layers. Their enlarged grain size alone produces this effect: When the size of component grains no more than doubles, the cohesiveness of a snow layer increases sixteenfold.

All such factors, and interactions between them, vary almost infinitely depending on countless nuances, and no one can guarantee whether or not avalanching will occur simply by adding this condition and that. Nonetheless certain rules of thumb are now well established, and the magic and divine supplications of previous days have given way to knowledge. One might think, for instance, that a well-cohered snow layer poses minimum risk of avalanches: It is holding together and should be secure. Actually, the opposite tends to be the case. Great slabs of snow lying like immense tiles on a roof can rocket downslope en masse in response to the least disturbance. Such conditions present the greatest danger. Loose snow, whether wet or dry, tends to stabilize itself by sliding before enormous pressures accumulate; but where slabs of compact snow are involved, avalanching comes as the final collapse of what may be an enormous mass. Sometimes with a sound no greater than a click, sometimes with a thunderous crack heard for miles, such snow breaks loose. On Mount Logan, Utah, observers once watched a massive avalanche drop more than a mile and then, as a cloud fifteen hundred feet thick, envelop a peak that stood high above surrounding terrain.

Slab avalanches of this sort often start out huge. A whole slope starts into motion virtually at once, leaving a low vertical face to mark its line of fracture. Each storm will have built its own distinctive snow layer with a crystalline structure and water content all its own. Each surface will have been tailored by weather conditions before fresh snow falls and builds an overlying layer. Sometimes a slick rind of ice from a rainstorm is sandwiched between an old layer and a new one, or the sugar-crystals of depth hoar act as ball bearings and ease the breaking away of a slab whose time has come. A layer may be ten inches thick, or ten feet; it doesn't matter. If it is cohesive it will slide as a unit. Loose-snow avalanches, on the other hand, occur when cohesion within the snow is slight. Individual grains start to slip at a point near the surface and, once moving, these grains sweep others along with them.

Wet snow or dry powdery snow can be involved in either type of avalanche. Wet-snow avalanches take place most often in spring, frequently starting near a sun-warmed cliff. Moving snow may ball up and produce boulder-sized lumps that keep rolling and sliding until they grow as big as houses. Rocks and soil and trees first get shoved ahead of such a hurtling snow mass, then engulfed in its plunge. Avalanches gather, swallow, and transport everything in their path—yet their speed seldom exceeds ten or fifteen miles per hour. Once halted, they turn to instant "concrete," often tens of feet deep. In the Alps there is record of snow from a wet avalanche so thick it took fourteen months to melt. In a separate instance a river flowed for a year and a half through an arch it had cut into a mammoth avalanche that was damming it.

Dry snow avalanches behave differently. They travel faster: Almost three hundred miles an hour has been clocked on a forty-degree slope in the Swiss Alps. Speeds are so great that dry snow becomes airborne as it avalanches. This recently was studied experimentally by spraying dye onto a snow slope, then deliberately releasing snow from above the colored area. The idea was to watch whether the avalanche would slide along the dyed surface or would churn and mix this snow in with it as it came. Gauges in the path of the onslaught showed that the avalanching snow slid until reaching a velocity of between fifty and seventy-five miles an hour, whereupon it became airborne and started to suck up additional snow from the surface.

Naturally occurring avalanches differ from those deliberately initiated in that they seldom behave quite so simply. Slabs usually start to disintegrate along the leading edge as soon as their speed reaches several feet per second. After that, fragments of what had started as a single slab begin to tumble and may whirl into the air and become airborne. Progressing downslope these powdery clouds disturb the air in front of them, and this whirls more loose snow into frenetic suspension. Speeds of nearly four hundred feet per second are known. Trees and buildings may be blasted by this wind and toppled before the snow hits them. There are cases of damage occurring after an avalanche has stopped; the dreadful wind alone packs all the devastating potential needed. Death has struck men untouched by the racing snow; yet, when autopsied, their lungs show lesions of the sort produced by explosions.

Today more avalanche tragedies befall mountain climbers and skiers than mountain residents. Not even falls kill as many climbers as avalanches do, and since skier's sport is wholly involved with snow, they are even more affected. Each weekend of the winter lowlanders fly from Los Angeles and London and Capetown to invade the Sun Valleys and Chamonix of the world. Some get easily drawn into a keyed-up excitement, even a bravura, yet have no particular mountain sense or snow experience beyond what a groomed slope feels like under fiber glass and metal skis. They thrill to the romance of the setting and the sport, accept the possibility of a broken leg, and know so little about snow dangers that the warnings of patrolmen may be resented or even ignored.

Others, who live at ski resorts and sustain their souls as well as their pocketbooks by cutting the new powder of one slope after another, may feel so familiar with certain avalanches that they venture into chutes knowing just how far and fast and deep the snow usually slides. Skiers playing this game get just enough into the chute to set the snow loose; then they ride the resulting avalanche until the snow begins to feel heavy, whereupon they glide out and back up on top of the snow to where it is still fairly light. This can be done only when conditions are precisely right, which is mostly in early spring. Try it on a day that is too cold, and the snow will be so dry it turns to a suffocating cloud as it avalanches, or too warm a day, and the snow behaves with impossible heaviness. For such derring-do a skier must know the peculiarities of the particular avalanche run he has chosen, and to learn it he must apprentice under a pro who has skied the area for years. At that, the game is reckless and likely to be deadly—which of course is much of the incentive.

Most skiers avoid avalanches so far as conditions and knowledge permit. Crossing a questionable slope they go one at a time, spaced a hundred feet apart. They avoid cornices, travel on the downhill side of crevasses as precaution against getting knocked into them by slides, and they stay away from the release zone of slopes. Moving along a ridge crest is safest, a valley floor second best. Generally, during the first part of the winter it is wisest to travel in the middle of the day whenever conditions threaten to set off avalanches. This best avoids the changes in temperature that are typical of early morning and late afternoon. Just the opposite holds true for late winter and spring, however. Skiing then is safest while temperatures are low, and the snow is firm. If travel is necessary during a blizzard it should be done in the early stages of the storm, before the new snowfall has built deep. The rest of the time it may be best to bivouac, even staying for a while after clearing has begun.

Skiers and climbers who have been caught by avalanches and survived describe similar experiences. "I heard the horrid hiss of snow and far above me the thundering of the foremost part of the avalanche," wrote a survivor who had been plucked from his feet while climbing above the Rhone Valley in 1864. "To prevent myself sinking in again I made use of my arms in much the same way as when swimming in a standing position. Then I saw the pieces of snow in front of me stop at some yards distance; the snow straight before me stopped, and I heard on a grand scale the same creaking sound that is made by a heavy cart passing over frozen snow in winter. I instantly threw up both arms to protect my head in case I should be again covered up.

"I had stopped but the snow behind me was still in motion; its pressure on my body was so strong I thought I should be crushed to death. I was covered up by the snow coming from behind me. My first impulse was to try and uncover my head—but this I could not do for the avalanche had frozen by pressure the moment it stopped, and I was frozen in." Luckily in this case the victim's hands still protruded above the snow, and although it took an ice ax to cut him free he was rescued swiftly. The sudden crushing pressure in a stopped avalanche can force all breath from the lungs, or can burst tires and cave in the roofs of buried cars, but this particular avalanche didn't provide pressure of those proportions as it stopped.

Best advice if swept into the seething morass of an avalanche is to "swim" with it, not against it, and to move toward the surface and the sides where velocity is least. Try to keep the head uphill and the mouth closed and either be standing upright or lying face down when the sliding stops. Both positions help avoid swallowing snow, which is dangerous because it may put water in the lungs and bring death by drowning. Hands held boxer-fashion in front of the face give a chance for having breathing space. Struggling is folly. It's too likely to cause sucking in snow and choking. Don't waste energy shouting unless voices can be heard close by, and even then don't shout if buried. Curiously enough, sound carries downward in snow quite readily, but not upward.

Survival odds vary. A miner buried by thirty feet of snow in 1966 miraculously clawed free by himself when an avalanche struck the Leduc copper camp on the fjord coast of northern British Columbia, while another survivor lay for three days entombed beneath only ten feet of snow before searchers found him. In the same tragedy—one of the worst avalanche disasters known—a Finnish carpenter named Einar Myllyla was swept fifty feet downslope along with the churned wreckage of his workshop. Rescuers scraped a helicopter into the snow directly above where he lay pinioned, able to move only his left arm. He could hear the clanking of bulldozers chugging back and forth five feet above him compacting snow, yet he felt no increase in the pressure against him. Later Myllyla heard voices and felt the vibration of helicopters landing and taking off, but still he could do nothing either to make his presence known or to ease his situation.

Of seventy-five men caught in the avalanche several still were missing as the fourth day dawned. Meticulous gridding of the whole camp area got underway, using the bulldozers to cut trenches while men rode the tops of the blades watching for signs of victims. As a dozer operator started across the helipad, a large slab of snow suddenly pulled free —and there lay the poor carpenter. He was looking up and welcomed his deliverers, asking them not to move him hastily because he thought his legs must be frozen, which they were. He had survived beneath the snow for three days and nights and after a short hospital stay in Ketchikan, Alaska, he recovered fully except for the loss of the toes on one foot and parts of three fingers from one hand. Ironically, two years later he was hit by a taxi in Vancouver and killed.

Perhaps the all-time record for mass burial by avalanche belongs to villagers from Blons, near Arlberg Pass in Austria. At the Leduc mine half the men in camp were buried. At Blons "only" a third of the population were caught but the total number was greater: 111 persons out of 376 in the village. Precautions had been taken. The ravages of snow were all too familiar. Each winter the village councilmen ordered removal of the crucifix that stood close to a certain ravine, lest it be damaged; and while crossing the bridge over that ravine villagers automatically walked in single file and quit talking whenever the snow seemed threatening. By spacing out they knew fewer would be taken should an avalanche roar down upon them. By not talking they knew it at least would not be their voices that set off the slope. Nonetheless on January 11, 1954, avalanches such as Blons never before had known broke loose, the first at 9:36 in the morning and a second at 7:00 in the evening. By the time the devastation was totaled, twenty-nine homes out of the ninety in the village were listed as demolished. Of the people who had been buried, thirty-three extricated themselves, thirty-one were dug out alive by rescuers, and forty-seven were found dead. Eight of those found alive later died, and two persons never were found.

One woman died from burns even though buried in snow. She had been baking when the avalanche hit her house and coals from the oven seared her as she was swept downslope. A man trapped for seventeen hours was alive when a search party got to him but died of shock when they told him how long he had been entombed. An elderly woman with broken bones and frozen legs was calm when rescued after twenty hours. Another woman trapped fifty-nine hours lived. So did a man trapped for sixty-two hours.

Few avalanche victims ever have survived longer than these people of Blons but one who did is Gerhard Greissegger, a young Austrian who in 1957 lay beneath avalanched snow for twelve and one-half days. He lost sixty-five pounds, yet lived. Even more astounding than this man's experience is that of an Italian woman and her two children, aged five and thirteen, who survived for more than a month in 1755 while trapped in a stable. Fortunately they shared it with assorted animals, and when a nanny goat gave birth to a kid the woman killed it so as to have the nanny's milk. A doctor who cared for the three wrote an account of the whole dramatic event titled, "A True and Particular Account of the Most Surprising and Happy Deliverance of Three Women buried 37 Days by a heavy Fall of Snow from the Mountains at the Village of Bergemoletto in Italy."

Such remarkable cases notwithstanding, the dividing line between a fair chance of rescue alive and very little such chance generally is about one hour; very few victims survive under snow for more than three hours. This is particularly true of skiers whose bodies almost always are directly encased in snow, as distinguished from victims partially protected by buildings or with some feeble source of warmth or food. A rule of thumb is that survival odds are halved by the passing of each thirty minutes. Surprisingly, a loss of consciousness and deep chilling sometimes actually prolongs life in situations where breathing is difficult. With a body temperature of 92° F., which is not much below the normal body temperature of 98.6° F., oxygen consumption stays at 80 percent of normal. At 86° the consumption drops to 70 percent, and at 74° it drops another 20 percent. Such hypothermia will lead to frozen arms and legs, but if the body-core temperature does not plummet too drastically life itself may be preserved.

The first rule for anyone present when an avalanche roars

downslope and swallows other members of a party is to mark where victims last were seen. Branches from a tree will do, or a ski pole—anything! But do the marking immediately, before shock compounds mental confusion and falling snow changes contours and disguises small landmarks. Next, quickly search below the point of disappearance for any sign of a buried person such as a protruding ski, or an arm, or some avalanche cord (which is bright colored nylon cordage worn trailing from the waist). Finally, send someone for help while the rest of the party continues to search. When found, an avalanche victim probably will be in shock, suffering from the effects of cold, and suffocating. There may well also be head, neck, back and/or internal injuries. Initial treatment should include opening the mouth, clearing it of snow, and employing mouth-to-mouth resuscitation if indicated.

Rescue systems are so well coordinated in resort areas nowadays that search teams expect to mobilize and get underway within fifteen minutes after learning of an avalanche. These first searchers form small teams of three to seven men trained in ski mountaineering and first aid. They carry little more than emergency medical supplies and lightweight shovels so as to travel fast. Led to the scene by a witness to the avalanche, they work fast. Their search involves moving in a line across the avalanched snow, each man scuffing the surface and scanning for clues. That failing, still in a line, the team uses its rods to probe coarsely along the fall line below the victim's last-seen position and also at the tip of the slide, in its eddies, along its sides, around any obstructions such as trees or boulders, and wherever abrupt changes in the slope may have altered the flow mechanics of the snow. The intent is to locate the victims quickly, checking the snow depths where the chances of survival are best. If possible more than one coarse-probe line may be in operation at the same time, or a single line may repeat its quick check an added time or two before going to the next stage of fine probing.

The fine probing is almost sure to locate the victim, but it is slow, and the chances of finding a body instead of a survivor are great. For the fine probe, men stand practically shoulder to shoulder with their feet about twenty inches apart. They gently thrust in their rods in unison, first by the right toe, then in the center between the feet, finally by the left toe. Everybody takes a short step ahead, and the process repeats. This methodically grids a slope every ten or twelve square inches. The rods are sections of steel or aluminum tubing ten to fifteen feet long and half an inch in diameter. They can be carried in short sections and fitted together at the site. Anyone who feels something within the snow leaves his probe in place, and the team leader, more experienced, makes additional tests. If shoveling is called for, care is taken to stop about a foot short of the detected object and change to digging with the hands. Harsh probing and frenzied use of shovels only add to the misfortune of victims. Success requires a sense of touch to feel the difference between rocks, frozen ground, wood, and human bodies. It takes an hour for twenty men to make a fine search of five hundred square meters of snow, probing to a depth of six feet. By using a coarser grid pattern, a greater area can be covered faster and victims' chances of survival shoot up. Unfortunately so do their chances of being overlooked.

The use of rods in avalanche rescue goes back at least two thousand years, as reported by Strabo. Travelers crossing the Caucasus carried staffs to push up through the snow should they have the bad luck to be buried by avalanche. At least this is Strabo's account, although probably the statement is reversed, and the staffs were used by survivors to push down and locate their trapped companions.

Dogs can effectively speed the search, especially those trained to sniff a human odor carried through snow. Saint Bernards were the forerunner of today's avalanche dogs, equipped for the task by size and strength rather than by any particular training. Furthermore, legend aside, they didn't carry flasks of brandy strapped to their necks. As a breed the Saint Bernards seem descended from Tibetan mastiffs brought to Greece in the fourth century B.C. and then on into western Europe by the Romans. Their fame— and name—comes from their association with monks at the

Saint Bernard hospice atop Saint Gotthard Pass. It was founded by a nobleman who decided against marriage on the very eve of his wedding, and instead joined the Augustine Order, building the hospice in A.D. 962 as an aid to travelers crossing the eight-thousand-foot pass.

Monks brought dogs to the hospice as defense against highwaymen. A few seem also to have been fitted with packs to carry milk from cowshed to kitchen, and it quickly became common practice for monks to take along dogs when they patrolled the pass for wayfarers. The dogs gave companionship, had a valuable sense of direction, and could easily plow through deep snow. They also could scent a man buried beneath snow. In this way dogs and monks became the world's first mountain rescue teams, extending help to all comers whether rich or poor, traveling alone or in groups. At Saint Gotthard and at other passes where the Augustinians also had monasteries, monks would go each winter morning to meet travelers approaching the passes from either direction. The constitution of the Saint Bernard hospice, dated 1436, directs this to be done from mid-November until the end of May "as the monks always have done," indication that the practice already was long established at the time the constitution was written. Rescues as such were not documented because recording them was considered vain. The number therefore never can be known, but during the early 1800s about fifteen thousand people crossed Saint Gotthard Pass yearly, and the hospice served up to four hundred meals a day. A note from that time comments: "Hardly a year passes in which one or more people do not die here; some are taken ill; others are surprised by the avalanches in the mountains, or are overcome by cold and freeze before arriving here, or being carried here."

The deliberate use of dogs specifically for avalanche rescue began by chance in 1938 when an avalanche buried a party of Swiss skiers. One had a pet terrier, and rescuers still searching for the last victim noticed the dog returning again and again to a spot where they already had probed. Its whining and barking finally led the men to reprobe there, and they found the skier still alive. News of this reached a dog expert in Berne and gave him the idea of specially training dogs to do what the loyal little terrier had done on its own. The Swiss Army picked up the idea and worked successfully with four German shepherd dogs, the breed most generally favored for avalanche work. Today dogs in Switzerland, Austria, Italy, Yugoslavia, Germany, Norway, Sweden, Canada, and the United States receive regular training in avalanche rescue.

The shorter the time a person has been buried and the nearer the surface he lies, the greater the dogs' chances of detecting him. A high permeability of the snow above a victim also helps; so does the temperature differential between the snow immediately surrounding him and that at the surface. A strong wind or false human scent will mislead the dogs. They need the first chance at checking an avalanched slope, before the hasty team of rescuers has trampled it. The best can make an overall search of two and one-half acres in thirty minutes, which is a fraction of the time men require. Given an hour or two it can complete a fine search. Any arousing of the dog's curiosity that prompts it to paw at the snow must be indulged until satisfied; then the dog can be urged on to investigate additional odors. Dogs used in rescue need to be intelligent, physically strong, have a well-developed olfactory sense, and be able to endure the intense light of open snowfields. Ideally each should be handled by his own trainer who is himself well versed in avalanche rescue. But if no trained dogs are available, any dog may prove valuable. The monks' Saint Bernards and the little Swiss terrier demonstrate this.

Various methods of locating victims have been tried other than searching with men and dogs. Accounts from as early as the seventeenth century mention clairvoyance. Another early system called for placing bowls of water along the sides of an avalanche, each with a piece of bread floating on top. The positions taken by the bread somehow were interpreted to point toward where victims lay trapped beneath the snow. Current efforts have involved trying to pick up a heartbeat or breathing, but so far snow's muffling effect has defeated tests of this approach. Devices to detect body heat also have failed, in their case because the snow of the avalanche itself is so churned that temperatures within it vary considerably. In a similar way, the differing densities of the snow have blocked attempts to locate a human by the body's effect on the gravitational field of a slope. Someday carbon dioxide sensors may work by picking up the faint traces of a person's breathing. Or experiments with butyric acid added to the polish used on ski boots may pay off. The acid is part of the odor of human skin and increasing the amount present for detection may raise the efficiency of avalanche dogs.

Some sort of X-ray fluorescence has been suggested but the hazards of exposure to such rays offset the practicability. Radar scanning has been tried from helicopters in the Alps. Magnetic detection also has been tested, in this case based on sensing anomalies in a magnetic field imposed on a slope. Placing a permanent magnet into ski boots might heighten effects and increase the feasibility of the method. For the present, however, electronic devices now on the market in America and Europe offer the best means of detection. A problem is that not all of them are standardized as to frequency, and they therefore are incompatible. United States Forest Service snow rangers and the International Commission for Alpine Rescue are pressing for standardization, however, and when this is achieved the chief remaining drawback will be human failure in letting batteries run down, or skiing without the switch in a proper "on" position. The devices work by generating a signal that can be picked up by other units set to the "receive" position.

To be considered successful any rescue method has to operate fast and with at least an 80 percent probability of success. It must be effective at a snow depth of ten or twelve feet, workable at low temperatures and in the midst of raging blizzards, be low in cost, and suitable for use over long distances and rugged terrain. So far men and dogs best meet these requirements, but modern wizardry is sure to augment them. Perhaps then the hazards of snow will be lessened.

CHAPTER

7

THE WINTER
BATTLE

NEARLY THIRTY YEARS BEFORE THE STAMPEDERS OF 1898 braved the passes leading to the Klondike, workmen were completing the main line of the Central Pacific Railroad (the first transcontinental railroad and a predecessor of today's Southern Pacific). It crossed the Sierra Nevada within the shelter of forty nearly continuous miles of wooden snowsheds designed to guard the track against avalanches. Gouged into canyon walls and snaking around curves, the run led through a dimness that was punctuated only by the even more total darkness of occasional tunnels. The sheds covered tracks, sidings, turntables, section houses, depots, and workmen's homes. Children played, and women cooked, cleaned, sewed, dreamed, and despaired in what Sunday supplements of the day labeled "the longest house in the world."

Laying track over California's mountainous spine nearly had defeated even the railroad greats Leland Stanford, Collis P. Huntington, Mark Hopkins, and Charles Crocker. The route led eastward from Sacramento to Summit, at sixty-nine hundred feet, where snow so tragically had held the Donner Party captive. From the pass the track twisted down the flanks of the mountains to Truckee and Reno—snow country practically every mile of the entire route. Yet in 1863 when construction started, engineers predicted a winter need for only a single short shed to protect one stretch of track and for two snowplows, one to clear track in each direction following a storm. The "Golden Spike" ceremony at Promontory, Utah, marked completion of the line coast to coast in May 1869; and flawless operation the following summer triumphantly crowned the long months of planning and working. Then came winter. The one shed gave the merest token of protection compared to what actually was needed. The two plows fought the deluging, drifting snow—and failed. V-shaped rams of wood mounted onto railroad wheels, they were powered by six to twelve puffing locomotives hitched together. If a drift was fluffy and light, the battering of the plows succeeded. If it was heavy they derailed.

Added to the shambles of that first winter's operation, an avalanche swept away one hundred feet of track and buried the right-of-way beneath a cementlike mass of snow, rock, and jackstrawed trees. The steam-powered plows, called "buckers," and scores of men with hand shovels tore into the blockade, but as fast as machines and muscles could clear the old, avalanched snow, wind piled up drifts of new-fallen snow. Trains from Truckee, beyond the mountains to the east, brought passengers as far as possible over the cleared line and from there railroad employees guided them on foot to Emigrant Gap, initially a seven-mile trudge. There they reboarded trains specially sent from Sacramento to carry them to the completion of their interrupted journey. Hundreds of passengers walked the decreasing distance between the two trains, while crews continued to battle against the intervening snow. Nobody seemed to opt for staying home and letting winter rule its mountain world unchallenged.

After that first chaos railroad officials decided to build more snowsheds. Woodsmen were scarce in California so Canadians were imported for the job. To supply them with timber, sawmills went up along the right-of-way. Through summer the sound of hammers tapped a message of progress as the peaked roof of a new, long shed straddled ever more of the track through the Sierra. Unfortunately, its "A" shape left space between structure and mountainside, and this void soon packed solid with snow. At first that seemed all right. Avalanches skated as intended over the track, their weight and force successfully borne by the combined snow wedge and shed. But officials rejoiced only until spring arrived. As thaw turned the packed snow from frozen state to liquid, a quagmire of mud was born. Timbers loosened, and since snow expands as it melts, pressure shoved sections of the shed out of line and even sent some crashing into the canyon below. Men tried to equalize the force against the two sides of the shed by shoveling snow from the uphill side to the downhill side, but it didn't work. The task was too gargantuan. Rebuilding offered the only hope, and as the thaw flattened parts of the old A sheds, crews used the timbers to start anew.

This time they built flat roofs which extended from above the track to the mountainside itself. That way snow couldn't pack against the shed walls. As further precaution four thousand shovelers, each paid $2.50 per day, struggled to ease the snow load on shed roofs and to clear unprotected stretches of track. The odds favoring successful winter operation began to shoot up, but summer passengers objected to having the view cut off by the sheds. In response officials ordered four-foot squares cut as windows, of course to be shuttered as necessary in winter. They also ordered planks in the shed walls to be set a few inches apart wherever hazard was minimal. This provided a flickering view and, incidentally, saved timber costs.

Thousands of people lived and labored within the snowshed. At stations spaced about every ten miles telegraphers operated track signals and passed along the orders that came over the wire for the train crews. Section foremen, housed with their families, bossed gangs of laborers who slept in bunkhouses and ate in cookhouses. The larger communities had a commissary car and housing cars connected to the main shed by covered walkways. Occasionally food ran low when trains got held up or cold became a problem with kerosene supplies exhausted and wood either waterlogged or wholly inaccessible beneath the snow. Children went to school from spring through fall and took vacation during the uncertain days of winter.

Men at work within the year-round dimness once in a while failed to hear the muffled rumble of a train coasting downslope and were injured or killed. Occasionally someone fell from the twenty-four-foot roof or was crushed by a boulder bounding from the peaks above and breaking through the roof. Explosions, collisions, and derailments added sporadic notes of horror which, at least once, were edged with comedy. This was in 1890 when a circus train jumped the track within the shed and its wooden cars burst open. Two miles farther down the line a lone gandy dancer was checking ties when he sensed a shadowy form approaching through the deep gloom. It looked like a lion! It came closer. There was the yellow shine in the eyes, the great mane, the swinging tail.

The man is reported to have run sobbing to the section house while the lion, equally frightened, fled in the opposite direction and crept back into its broken cage amid the train wreckage. Life then settled back to routine—except for monkeys which hid for weeks in the dim nooks of "the longest house in the world."

The worst single avalanche disaster in the United States took place in 1910, not in the Sierra Nevada but in the Cascade Mountains of Washington. It involved two trains on the Great Northern line, both of them westbound from Spokane for Puget Sound. A massive avalanche blocked first a passenger train, then a mail train. Consequently on the morning of February 23, snowbound men, women, and children trooped to a nearby railroad cookshack for breakfast. Snowflakes "the size of soda crackers" whitened heads and shoulders as they walked, and only the prospect of hot food eased exasperation over the delay. Great Northern had completed its route in 1893, a quarter century after Central Pacific, and it was the last of the historic links through the western mountain barrier. In the years since completion more than four thousand trains had made winter crossings, none delayed over twenty-four hours. Officials felt proud of their company's record and expected this latest challenge to do nothing to blemish it. Fate decreed otherwise.

Men fought two days and nights to break through the slide and the newly forming drifts, and as soon as success seemed imminent the superintendent of the Cascade Division ordered the two stalled trains brought to Wellington, just west of Stevens Pass, to await the final unplugging of the track. Parallel sidings could accommodate them there, and the hotel would greatly increase comfort for the stranded passengers. The plan had merit, but the snow refused to surrender. First a plow broke down while chewing into the slide. Then, before its replacement arrived, one fresh avalanche covered the track, and a second demolished the cookshack that had served as emergency restaurant. The cook and his helper were killed. Passengers realized they too might have perished if the train hadn't been moved, but the news made them look all the more apprehensively at the white slope poised above their new position. Some urged backing the train into Cascade Tunnel, just upslope from Wellington, but railroad officials refused because they feared that doing so would risk asphyxiating the passengers. With the locomotive continually fired to heat the coaches, its fumes seemed a greater risk than an avalanche, which might not happen.

Snow fell at the alarming rate of a foot an hour, and word of several avalanches along the line reached the marooned passengers and workmen, but their own slope was holding. By the afternoon of February 26 the plow fighting the original blockage became trapped as snow roared downslope behind it. Twice within the past four days crews had broken through, but the snow's fury never eased enough to let the trains pass. The slopes repeatedly unleashed new avalanches, and on March 1 when falling snow turned to rain, the situation worsened still more. The rain increased the snow's weight. A snow slab estimated half a mile long by a quarter mile wide and twenty feet deep roared into motion. The passenger train, the mail train, several box cars and electric engines, railroad men's private cars, and a rotary plow all plunged over the cliff to lie beneath tons of snow. A few victims were spewed out onto the surface of the snow and survived. A very few of those buried by the snow were dug free still alive. Most perished. Ninety-six of the people aboard the trains died. Twenty-two lived. On March 12, eighteen days after the first plow had gone into battle against the first avalanche, the track finally was forced open enough for a train to roll.

Forty miles of track had to be abandoned as a result of the disaster. This section was rerouted in a tunnel blasted through the mountains for eight miles.

Trouble of this sort in the Cascades scarcely was Great Northern's introduction to the horrors of winter, however. The infamous north plains blizzard of 1887 had seen to that. In places coal and hay supplies ran critically short, and mail was delivered on snowshoes. Newspapers in the region, such as the Livingston (Montana) *Enterprise*, carried blizzard and railway stories from early January of that year into mid-March:

January 8. The west-bound trains on Saturday, Sunday, and Monday were from 2 to 3 hours late . . . caused by heavy snows in Dakota.

January 22. The railway Co. has met with more than usual trouble this winter by reason of snow drifting on the track about the depot. [This despite snow fences and a plow every few hours.]

January 29. . . . snow drifted so solid the engine several times left the track. . . . trainmen camped 12 mi. from Livingston & the conductor walked for help.

February 5. No such difficulty has been experienced in operating the Northern Pacific railway since its advent into the country . . . yet no expense or effort has been spared.

February 12. Kind-hearted people of Butte subscribed a liberal sum of money and expended it for provisions for relief of emigrant passengers on the snowbound train at Harrison last week, who were 36 hours without food.

February 26. People in other sections have no conception of the fearful cold prevailing. . . . Trains operate with the greatest irregularity and all other roads are lost. . . . The people are burning wheat and corn . . . and household furniture. . . . Blizzards drive fine snow into houses. Many persons have frozen to death—more than for a quarter of a century past.

March 12. A train was run on the [Yellowstone] Park branch Sunday, the first since the one which rescued the Schwatka party six weeks ago.

Ironically, that same March railroad men successfully tested rotary plows at Salt Lake. They were an altogether new concept and had they been available even a few weeks earlier they quite probably would have eased the anguish of north-plains railway operation during the record blizzard. At Wellington thirteen years later they failed, but there mountain avalanches outflanked and overpowered them. On the plains, against lighter snow, they probably would have sufficed.

The whole audacious idea of overcoming winter snow was new as the twentieth century opened. Until then men had accepted submission to nature as their lot. Nor was this necessarily negative. Snow, far from an impediment, was utilized. Winter's blanket eased the problems of getting around. Webbed snowshoes in America and wooden skis across Eurasia made winter by far the easiest time of the year for travel. Snow smoothed terrain and opened limitless routes apart from established trails and waterways. Furthermore, it rendered the tracks of game animals visible even to inexperienced eyes and silhouetted the beasts themselves for marksmen of whatever skill. As modern living patterns took over from indigenous patterns men still saw snow as a transportation ally, not an enemy. Villagers of the European Alps kept travel routes open by hitching six or so horses—

or in some areas cattle—to an empty sledge. A heavy chain dragged from the back to cut the path through new-fallen snow, and additional horses followed behind with empty sledges to firm the course. Men on foot added finishing touches with shovels. All day might go into opening five or six miles of road. If snow lay deep and wet, the lead horse had to be changed every half hour, but in that era there was time to spare and being snowbound generally gave a feeling of coziness rather than of emergency. Businessmen didn't feel cut off from the airport, or housewives from the supermarket, because life didn't yet offer such blandishments and the human sense of pace wasn't at odds with nature's changing moods. Such winter travel as there was depended on runners, not wheels, and men groomed snow to ease their glide rather than pushing it aside or hauling it off.

In Canada and in northern United States road departments developed rollers to pack snow in place and provide smooth roadways. Shoveling or plowing snow off the road left a trench that was likely to blow shut or at least close in so narrow that teams scarcely could pass. Rollers worked much better. They provided a wide road easy and cheap to maintain partly because instead of being sunken it stood high and exposed, and fresh snow tended to blow off. Also packing the snow throughout the winter extended the sleigh season into spring, thereby shortening the weeks of sloppy travel through mud. In some cases winter crews used plank drags to supplement rollers, but overall the rollers did the job. They were enormous. Most were of wood, three to six feet in diameter and twice that long. They weighed two tons or more. Sometimes two drums were mounted together on a frame of oak and topped with an operator's seat and toolbox. Roller surfaces usually were oak too, fashioned of staves four inches square which were held by iron hoops.

The coming of World War I brought the need for War Department trucks to keep rolling regardless of winter, and and to accommodate them horse-drawn blades and motor plows began to gain favor over the rollers. Snow needed to be cleared, not compacted, and municipal street departments, as well as their rural counterparts, switched from packing and smoothing to shoveling and plowing. The yearly "Report of the Commissioner of Street Cleaning for New York City" gives a running account of man's growing battle against snow. The 1901 report tells of 843,964 cubic yards of snow removed from city streets at a cost of thirty-seven cents per yard. Men hand-shoveled the snow and hauled it to the river to be dumped. By 1905 the commissioner pointed with pleasure to increased efficiency and a drop to fourteen cents per yard for snow removal. In 1909 snow-plows were introduced, an "innovation so successful" that the Board of Aldermen granted $10,000 to purchase additional plows. The 1911 news is sewering: "The Bureau of Sewers in the Borough of Manhattan allowed the use of a number of manholes [for snow disposal]. It was found . . . without detriment, and it is hoped that a more liberal policy in this regard will be adapted. . . . A number of protests were made against dumping snow in vacant lots in the Boroughs of the Bronx and Brooklyn, and I trust that citizens will use judgement in raising such arguments, as the cutting off of these relief dumps will impede the work of snow removal and add greatly to its cost."

Today the Snow Removal Manual for New York is half an inch thick and covers topics from The Duties of District Superintendents (and also those of Foremen, Clerks, Time-keepers, and Squad Leaders) to The Registration of Emergency Truckmen and How to Keep Motor Vehicles in Readiness for Snow Removal Operations. A supplement covers strategy. It stipulates that snow is to be fed into the sewer whenever possible, taking care not to choke the system by overloading or letting debris get mixed in. If the temperature rises above freezing, snow should be "scattered" by pushing it from piles and ridges into the path of oncoming vehicles which churn it to slush and melt it; if below freezing weather is forecast the operation reverts to "piling," taking care to keep crosswalks and bus stops clear. "Hauling" takes over as a last resort and "Approved Waterfront Locations for Disposal of Snow" are compiled each year.

The cost? Even if not a flake fell the city would have to

be prepared for the worst. The price has gone as high as twenty million dollars in a year (for 1960–61 when storms dumped fifty-seven inches of snow). In ordinary years most of the money is spent on rock salt—over one hundred thousand tons of it. In addition an inventory of automotive spare parts has to be maintained to guarantee speedy repairs (cost, around one quarter million dollars) and the equipment itself carries a high price tag and a high amortization rate. Nearly one thousand vehicles are needed for snow removal, about 20 percent of the department's entire fleet. Personnel engaged in the winter battle number more than ten thousand on a regular basis and double that in emergencies.

Eight million New York residents and about the same number of commuters look to the Sanitation Department to underwrite their chances to roll freely all winter. The letters file at department headquarters bulges. Brides write that they phoned for a plow and thanks to it managed to get to the church on time. Parish priests comment that "the basketball lads had their tournament after all." Market district merchants point out that food shortages follow when snow problems aren't well handled, and one recent letter "in behalf of the Association of Poultry Slaughterhouse Operators, Inc." mentions a particular Monday morning during a storm "when over one million pounds of live poultry—a highly perishable food product—was expected to arrive by truck. . . . To the amazement of everyone, the Terminal yard was found free of any obstruction and traffic moved normally to the market. Thanks."

At times despite sophisticated mechanisms and all-out human effort, winter fury dominates. It happens even during years of fairly routine snowfall and temperature, as for example 1969 in New York. Airports shut down in early February that year, stranding six thousand passengers. Long Island Railroad suspended service. The Transit Authority openly described bus and subway transport as "very uncertain," and auto traffic as hopeless. "It isn't much of a Fun City," reported the New York *Daily News* on its front page for February 10. Power lines had snapped, overloaded with snow. The Health Commissioner had warned anyone with

heart or lung trouble to "leave the shoveling to somebody
else." Schools were closed and garbage collection halted
because every possible man in the Sanitation Department
had joined the battle against the snow. The stock market
shut down, and this alone cost the city an estimated three
million dollars in the loss of transfer receipts and sales taxes.

Drifts as much as six feet deep blocked streets, and fire-
men struggled on foot to meet calls. Fathers with telephones
delivered babies at home following advice from hospitals,
and those stranded in cars with their wives aided the births
on their own. Lincoln Tunnel was closed, and one thousand
autos were stalled on the Tappan Zee Bridge, their occupants housed at toll stations and police barracks. Jet planes
moored in their berths at Kennedy Airport offered continu-
ous-run movies to stranded passengers, and when a heart-
attack victim newly released from a hospital suffered a re-
lapse he was treated in a hangar by a cardiologist, also
among the stranded.

Such is uncontrolled snow in New York: lives in jeopardy,
financial disarray, inconvenience, and irritation. Twenty-five
inches of snow fell that winter of 1969, and the city spent
almost eight million dollars to plow, shovel, pile, haul, and
melt it. On the whole it was an average situation except for
the week of havoc which came about largely because the
Sanitation Department held back as the storm began and
then couldn't quite catch up. Optimum success in the battle
against snow hinges on correctly gauging the time to mount
full assault. Rushing needlessly violates financial responsibility. Delaying past a critical—and unpredictable—point
denies citizens their mobility.

Multiply the problem of New York's efforts through a
winter by all the streets and roads and railways and airports
of the world's snow belt, and modern man's winter mobility
becomes an immense burden in terms of money, hours,
square miles scraped, and tons of snow pushed, blown,
hauled, melted—and cursed. A few years ago the U.S. Army
Snow, Ice, and Permafrost Research Establishment compiled
a bibliography of techniques for solving the snow problem.

They surveyed nine air bases, including one in Canada, thirteen state highway departments (California, Colorado, Maine, Massachusetts, Michigan, Minnesota, Montana, New Hampshire, New York, North Dakota, Vermont, Washington, and Wisconsin), and six European nations. They found 370 patents for devices to remove snow and ice, also that no systematic program coordinates research on snow removal. Instead the development of methods and machines has come largely through trial and error, which in part can be expected since winter circumstances vary so greatly from region to region. Good ideas can't apply universally. In the mountain passes snow falls often, and it piles ten to thirty feet deep. Avalanches churn rocks and trees with the snow they bring down. Grades are steep and curves sharp. Operators and equipment face problems far different from those on the plains where drifting provides the key challenge.

In a state like Washington all types of snow must be dealt with. On the west slopes of the Cascades wet snow from ocean-born clouds buries slopes. On the plateau east of the range a much lesser amount of dry snow conspires with wind to form drifts sometimes so hard-packed that plows bucking into them ricochet off. Along the Columbia River graupel rolls down the sides of the gorge onto the highway and the railroad track. Rain may saturate any of these snow types and highway men then must deal with weights of half a ton per cubic yard. Thaw will cover the road with meltwater that is likely to freeze into a death-dealing glaze, or, seeping under the pavement, the melt leads to jolting chuck holes.

Ice causes the worst removal headaches. Even where snowfall is heavy and troublesome the cost of getting rid of it runs a tenth or less the cost of removing ice. Unless the temperature hovers near freezing, ice can't be handled mechanically; the bottom half inch will be bonded so tightly to the pavement that scraping it damages the road surface. Some highway departments use a serrated blade to groove the ice (or packed snow) and increase effectiveness, and Minnesota has experimented with a ribbed roller for crushing ice. It is mounted on the rear of a heavy truck and loaded hydraulically to deliver more than a ton of pressure per square inch. Unfortunately it operates too slowly to be economical, and, worse, it often breaks the pavement along with the ice.

Ice presents such problems that highway and street departments struggle to eliminate snow before it compacts or melts and refreezes. Salt once appeared a cure-all but alarming side effects give reason for considerable disrepute. Snow control in the United States alone consumes one-sixth of the world's total supply of salt. This represents ten million tons applied annually to sidewalks, streets, highways, and byways. The use averages from four hundred to twelve hundred pounds of salt per mile, per application. Cumulatively through a winter many roads receive more than twenty tons per lane mile.

This reliance began after World War II with acceptance of a bare-pavement policy. Previously traction on snowy surfaces simply had been enhanced by spreading abrasives such as sand and cinder, but salt seemed to promise a better way. In the first years its use totaled about half a million tons for the whole nation—enough to demonstrate effectiveness and spark such acclaim that the volume applied has doubled every five or six years since. Road salting now amounts to a 150-million-dollar business, with Pennsylvania, Ohio, New York, Michigan, and Minnesota leading the way. Salt speeds the melting of snow and ice, weakens the bond to pavement, and by lowering the freezing point minimizes the formation of new ice. It turns two or three inches of snow, or an ice crust, into heavy slush that passing traffic splashes off onto the sides of the road. Cheapest and most widely used is rock salt, sodium chloride. Calcium chloride is the next most commonly used. Its cost is triple or quadruple that of sodium chloride, but it has considerable advantage because it works faster and stays effective at lower temperatures.

Damage to road surfaces became noticeable as soon as heavy salting began. Concrete needs multiple coats of boiled linseed oil and petroleum mineral spirit or kerosene to pro-

tect it from salt. Asphalt seems less affected, although baring
pavement instead of leaving it insulated by snow increases
damages from frost heaving. Corrosion of automobiles
and trucks owing to salt costs an average of sixty dollars per vehicle per year in snowbelt states. Along road-
sides excess salt damages plants as drastically as if drought
had struck them. Wholesale scorching, stunting, and de-
foliating are part of the cost of present-day snow-free pavement—a high price in dollars as well as in amenity loss
and nuisance. New England towns typically lose fifty to
one hundred mature roadside trees per town per year, each
valued at from one thousand to five thousand dollars according to insurance claims.

Tests show salt damage as far as one hundred feet from
the edge of pavement. It harms soil structure and by up-
setting osmotic balance causes water to be drawn out of
plants' roots instead of into them. (The flow is toward the
greater salt concentration, which normally is in the root
sap.) Gardeners at Arlington National Cemetery have reported the loss of privet hedges and bluegrass lawns because
of salt, and at Walter Reed Hospital street maples and parking lot hemlocks have died because of salt-laden snow piled
outside curbings where its melt seeped into the ground.
In New Hampshire, Massachusetts, Vermont, and Connecticut tests of afflicted and dying trees indicate excess sodium
in leaf tissue and a high total salt concentration in sap.
Maples seem the most affected, sugar maples more so than
red maples. The New Hampshire tests noted damage along
state roads, which are salted, but not along roads of the
same area where de-icing salts are not used.

Animal and human life suffers too. The deaths of pheas-
ants, pigeons, quail, and rabbits have been traced to salt
poisoning in the Madison, Wisconsin, region. Deer drawn
to roads to lick salt have been struck by cars. Dog owners
protest damage to pets' feet. Doctors in several areas of
high winter salt use recommend the purchase of distilled
water by patients with congestive heart failure, certain
kidney problems, or hypertension, and for many pregnant
women. Widespread contamination of municipal wells has

become commonplace and a few communities have been forced to close water supplies that had served for generations. Some of the trouble comes from snow scraped off streets and roads and dumped into rivers, lakes, and ponds; some is from seepage through the soil. Improper storage causes a high proportion of the problem. Salt piles should rest on an impermeable footing and be covered as protection from rain and snow, but often no effort at containment is made.

The peak of the trouble comes in January and February, when salting is heaviest. Where new highways provide a virgin situation for observation, the effects of salt can be seen to increase with each passing winter. Many are only beginning to be recognized. For example sodium in streams and ponds favors the growth of blue-green algae, which upsets the biotic balance and in time can harm water quality. Both sodium and calcium have the potential of interacting with mercury, which often is present in mud pond bottoms, and the combination can release deadly toxin into the water. Other heavy metals probably also are released by the same sort of ion exchange. Additives used with salt to reduce caking threaten still more trouble. Dissolved in water, the most common of these substances (ferrocyanides) can generate cyanide in the presence of sunlight. In a Wisconsin study, water moderately contaminated with highway de-icing compounds was found to produce in just thirty minutes nearly four times the cyanide level permissible for public water.

What are the alternatives to salt? Burlington, Massachusetts, recently returned to plowing and sanding, and reports an immediate 22 percent lowering in the cost of snow removal. The figure can't mean much until it represents more years, but it may be a harbinger. Oregon never switched from abrasives to a primary reliance on salt. Their system is to mix just enough salt into sand or cinders to ease loading and handling and to speed their imbedding into highway ice. This amounts to about fifty pounds of salt per cubic yard of sand. Until the late 1940s abrasives were the principal way of fighting icy pavement in North America, Europe, and Asia; but since then, particularly in the United States and Canada, more cars and faster driving had reduced the effectiveness of that method even before using salt became popular. The passing of fifteen or twenty cars, or trucks, tends to blow sand aside. Wind does so even more drastically. The loss can be lessened, however, by applying sand heated enough to melt into the snow or ice surface and freeze in place, no longer susceptible to passing cars and gusts. Using damp sand or following the sander with a spray truck accomplishes the same thing if the temperature is freezing or below. Also, if dark abrasives are used they will absorb heat and melt into the surface, then freeze into position.

In regions where drift causes the greatest problems, road maintenance headaches can be minimized by building snow fences, baffles, and walls. How much snow the wind can carry depends on the cube of its velocity and any lull may cause it to drop its load. Therefore if the position of obstacles is deliberately manipulated, the formation of drifts can be controlled. A solid wall will produce an abrupt deposit mounded against and over it; a fence with slats as far apart as they are wide will form a longer and flatter drift. Three quarters of it will lie in the lee of the fence, and the size of the drift will depend on the height and slant of the fence and the nature of the opening between it and the ground. By guiding the wind, baffles can cause snow to sweep across a road and leave it clear, or force the deposition of snow before it reaches the road, or other predetermined area. Trees can be planted for half the cost of putting up and taking down fences, and in five to ten years they usually outperform a fence. Makeshift wind fences made of blocks of snow and also of fir boughs were used in Russia and Germany during World War II. In Denmark fences of straw rope, and in Japan of woven reeds, still are respected over wooden-slat fences of equal density; the roughness produces multiple baffles which foster snow deposition.

Heat is used to melt snow from critical short sections of pavement such as toll plazas and bridge ramps, but it is too expensive to warrant general use. Heat sources vary from overhead infrared lamps to electrically charged wire mesh imbedded in the pavement. The Road Research Laboratory in England is experimenting with graphite and other electrically conductive substances mixed into asphalt to permit heating entire road surfaces. In parts of Canada ethylene glycol (antifreeze) is circulated through a grid of pipes set into the pavement. At Klamath Falls, Oregon, the heat for keeping streets clear is geothermal. Water is piped into a hot zone within the earth, and steam is piped back up and into a grid beneath the streets. Mountain villages in Japan also use geothermal heat to control snow, but they spray the hot water directly onto the streets, flushing away the resulting melt in open ditches. A Russian system taps into the steam heating of buildings, using a hose to connect a street-level network of perforated pipes to rooftop steam vents. Several nations use mobile heat to melt snow. Machines mounted on trucks suck in snow, partly melt it, filter it to eliminate debris, then pass it into a hot-water tank to finish melting and be drained through a hose that trails behind the truck.

Researchers throughout the world's snow belt are trying to develop additional approaches to the winter battle. Perhaps a substance that absorbs solar energy could be imbedded into pavement. Maybe a Teflon coating would prevent snow or ice from bonding to a road surface, or an environmentally safe and inexpensive chemical could be applied to weaken the bond. Urea and glycol are known to be effective but are too expensive. Maybe snowplows can be equipped with jets of compressed air or water to break ice crusts and simplify scraping roads, or improved tire designs might increase automobiles' traction enough that winter highways would need less attention. Whatever the odds, man seems intent on continuing to roll. The days of surrendering to snow are past.

CHAPTER

8

SNOW AND LIFE

SNOW HAS TO BE SHOVELED. IT MAKES DRIVING DANGEROUS. Avalanches bring sudden death. Blizzards kill cattle. Say the word *snow* and modern man reacts negatively. But the tie between snow and life is ancient and not wholly malevolent.

Most insects time their cycles to be in dormant, pupal phases while snow mantles the land and food is unavailable, yet ladybugs by the millions have been found hibernating beneath snow. Birds generally migrate, but some species are superbly adapted to life in a cold and snowy world. Ptarmigan change their plumage to camouflage white and grow special feathery tufts on their feet as snowshoes. These feathers even fold out of the way as the birds bring their feet forward, then flatten on the downstroke to increase the bearing surface of the foot. For insulation from winter's cold, ptarmigan dive within snow's protective blanket. Sometimes they plummet into it from a tree branch; other times, after feeding on willow buds, they simply continue walking, gently angling down into the snow. Grouse similarly snuggle beneath snow. Each winter they grow special spurs that help them dig their way in.

For some mammals snow provides the only usable source of water in winter. Gazelles in the Gobi Desert rely on *snow mines* buried beneath sand. These are drifts of snow blown into hollows and covered—and insulated—by sand a foot or two deep. If it weren't for digging to them, gazelles would die of thirst. Arctic and subarctic musk oxen also depend on snow for water. They are unusual in seeming to have evolved a means of recycling nitrogen within the body instead of discharging it in urine. More study is needed to know positively but, if true, this appears to be a direct physiological response to snow. Snow provides water but is not an ideal source. A great deal must be swallowed to get a little liquid, and in the process the body gets chilled. Consequently, to reduce the amount they must eat, musk oxen concentrate nitrogenous wastes within the body rather than eliminating them in urine and moist fecal pellets. Several desert creatures similarly regulate the nitrogen in their bodies in order to survive with a minimum of water. Musk oxen, however, do so only during the winter when ample water is unavailable. All of life's balances rest on an especially delicate fulcrum then, and even such small savings in body heat can be crucial to life.

A road through snow country may spell death for caribou; wolves benefit from snowmobiles; early snow covering lake ice helps muskrats: Such were the observations of a trapper my husband and I met while exploring back roads along the border between British Columbia and Yukon Territory. He was Gunther Lishy, a soft-spoken, rosy-cheeked, infinitely strong, and gentlemanly emigrant to Canada. Isolation was his antidote for having been taken prisoner of war in Russia during World War II, and in the bush trapping and hunting often are the best way to make a living.

"They get you your bread but don't put much butter on it," was Gunther's assessment. "Life is for the living alone, not for anything much beyond that."

The work has changed since his arrival in the Canadian outback in the late 1940s. Anybody can be a weekend trapper now. Roads and snowmobiles make it possible, and they directly affect the animals as well as facilitate the trappers' travel. "A caribou can wander a road in winter, then hear a snowmobile coming and have to step off all of a sudden, and it'll sink to its belly in soft snow alongside the road." This happens mostly in spring when there is just enough sunshine to put a weak crust on the snow. Caribou can't move easily when they break through, and a man on snowshoes can outrun one in a quarter mile.

The effects of snowmobiles are "too many to count." Their tracks give wolves highways of packed snow to use in overtaking moose or deer. Sharp hooves break through the snow of the tracks; broad, soft paws don't. Trappers use the machines' tracks to advantage too. "You can get otter that way. You break trail with a snowmobile, or even snowshoes, deliberately crossing bends in a river. That makes shortcuts for the otters. Their legs are so short they sink to their bellies in soft snow, but compacted snow is easy for them so they'll follow your track if it goes anywhere near where they want to go. All you have to do is pack a trail and set a trap.

"Muskrats, now, they benefit from a deep snow in autumn when freeze-up is just starting. A good snow blanket keeps lakes and rivers from freezing solid and the muskrats can keep on swimming around and feeding right into winter."

Humans—even those who live in cities and those who escape to Caribbean islands in winter—depend on snow. It grows crops to feed the cities and turns the turbines of industry, which provide the dividends that permit the escapes. Clouds are seeded to increase winter snow accumulation, and regions argue over who rightfully "owns" the clouds and has the right to seed them and so increase their mountain snowpack at the expense of others elsewhere. Even forests are being tailored to manage the snowfields beneath the trees for their water content. This works because the fewer the trees the more the snow that reaches the forest floor, with different kinds of trees producing different results. Nobody yet understands exactly how it works, however. Serious investigation of snow so far has concentrated more on indexing snowpacks already on the ground than in detailing their formation.

The type of snow falling and the temperature affect what clings to trees and what sifts through. At the base of Mount Rainier, where below-zero temperatures are rare and snow is lovingly dubbed "Cascades cement," the hemlock and Douglas fir beyond our window turned to giant plumes of white practically with each snowstorm. Even fences became snow sculptures. In the Badlands of North Dakota, however, the cottonwood trees beside our cabin almost never were whitened with snow. Not even the junipers growing in the ravines held much snow. Dakota snow is cold and dry and less likely to stick than the relatively warm, moist snow of Pacific coast regions.

Only occasional direct observation has been made of how snow builds in tree canopies. In a Colorado forest of lodgepole pine and Engelmann spruce, snow watchers found that some crystals hit needles and slide down, and some ricochet. Once they begin to stick, the rate of accumulation accelerates; but by the time ten or fifteen particles have lodged against a needle, wind usually knocks them all away. If it doesn't, incoming snow will stick cohesively and perch on top of needle clusters rather than really anchor within them. Pine needles start trapping snow earlier than spruce needles do, probably because they are longer and also because pine branches grow more upright than spruce branches. Even so, by the end of each storm the two species seem to hold the same amount of snow. If anything, the denser growth of spruce is slightly more snow-laden than pine despite its later start in accumulation. No actual measurements were made during the storms observed, and such a lack of precise information typifies the gaps in a full understanding of snow and trees. Nobody yet knows enough about the specifics of what happens.

Once snow has caught in branches does it simply fall to

the ground in time? Is wind required? Tests in Germany using the rotor blast of a helicopter showed that wind cleared snow from the top ten feet of trees without affecting the snow on lower branches. In other experiments shaking trees seemed to remove only about two-thirds of their snow.

What if snow doesn't fall from the branches but instead melts and trickles down a succession of stems to the ground? In this case its moisture will enter the ground water regime more quickly than that from the snowpack itself. Or, if instead of trickling down, the melt drips directly from the branches it will sink into the snow on the ground and thereby will increase the snow's density and tend to speed its melt.

Some snow neither falls to the ground nor melts. It sublimates: evaporates directly from solid state to gaseous without passing through a liquid state. Consider, for example, snow held in trees. Branches receive more incoming radiation than the ground beneath them does, and they also receive radiation reflected back from the snow surface and heat given off from the earth. These differences in radiation affect sublimation. Pans filled with snow and placed both on the ground and in trees show a sublimation rate within the trees that is triple that at ground level. But nobody knows precisely what happens to vapor sublimated from branches. It may vanish into the atmosphere, condense as rime onto branches, or diffuse through the trees and condense onto the surface of the snowpack.

Trees markedly affect the buildup of snow. So, too, the buildup affects the trees and other forms of life. Let a spruce in the northern taiga tip slightly off vertical, and it will catch more snow than those still standing upright. Eventually the load breaks the tree or causes it to fall, leaving an opening within the forest. Other trees ringing the opening then grow most luxuriantly on their sides facing the extra light. In time the weight of snow caught on their branches overweights and breaks them, and the opening grows. Shade and the rain of dry needles from the branches are gone, which soon affects the plant community. Mosses die out and willow, alder, aspen, and birch take over. In time, snow bends these new branches and brings their tips within reach of snowshoe hares and ptarmigan. The surface of the glade becomes hummocky with caves formed beneath snow-weighted branches, and there small creatures take refuge. Nitrogen from their fecal pellets enriches the soil, and eventually spruce seedlings gain a new start. The succession of plants has come full circle.

Human understanding of the fine tuning of such relationships between snow and life has only barely begun. The elk of Yellowstone National Park, Wyoming, furnish an example. Massive winter die-offs during severely snowy years are normal for herd animals, but not even biologists have known or accepted this until recent decades. Judging from fossils, elk, also called wapiti, have been roaming the mile-high Yellowstone country for at least twenty-five thousand years. They are a North American species that migrated over the Bering land bridge during the last ice age. In time they ranged the continent from coast to coast and from the far north of Canada practically to Mexico. By early in this century they had decreased so that only about seventy thousand were left, perhaps half of them in the Yellowstone region. Establishment of the national park in 1872—the world's first such preserve—gave the elk sanctuary at what probably was a crucial time in their survival as a species. Man's knowledge of how to help them was faulty, however. The urge—and need—to protect domestic livestock extended to management of the wild elk herds. In those early years all predators from wolves and coyotes to cougars and grizzlies were categorically branded as "bad," and systematically poisoned, trapped, and shot. As a result elk herds within the park increased. So did winter die-offs.

The only correction seemed to be for firing lines of hunters to form along the north boundary of the park each fall and shoot elk migrating out of the snowed-in high country toward their traditional wintering grounds at lower elevations around Gardiner. Park managers and the public believed the die-offs came because the elk herd was too large. More accurately the problem lay in men having usurped land the animals needed for winter survival and having killed the predators that in nature's scheme help to check burgeoning overpopulation. In fact, the "problem" wasn't a problem except in men's understandable reaction to the die-offs as cruel and wasteful of life. Actually they were normal. Die-offs forcefully assure herd turnover, hard on some individual animals but an advantage to the herd in the long run. Shooting prime animals for freezer and trophy wall, on the other hand, is hard on both individuals and the species—hard on the well-being both of individuals and of the species.

Today's elk management in Yellowstone has become a planned nonmanagement. The fall firing line is not permitted, and the practice of live-shipping elk out of the park, which briefly replaced hunting as a control measure, also has been discontinued. Herd size now is maintained as it was through the long millennia before men tried to alter the system. Predators have been allowed to return to the park, and winter die-offs are accepted as a natural result of heavy snow. During the winter of 1969–70, when snow came late and lasted long, one of the park's elk herds lost perhaps a quarter of its number. Most deaths were among first-year calves. The cause was malnutrition, and their loss was of no consequence to herd well-being except possibly as an asset. The strongest calves lived, and theirs were the best genes to pass on in any case. The dead, weak yearlings would be replaced by spring's births.

Second hardest hit by the deep snow were the bulls that had bred the most actively during the previous rut. Harems take time! The male stalwarts of the rut are so busy with cows that they don't feed well. They enter winter with poor fat reserves. Yet eliminating them also is an advantage. It assures different bulls and different genes when the next rut comes. As for the dead themselves, their flesh gives life to the community as a whole. Creatures from bears, wolves, coyotes, foxes, shrews, and mice to ravens, jays, magpies, and chickadees depend on such deaths for their own lives. Two days after an elk falls, every trace of its carcass will have disappeared except perhaps for a frozen bloodstain in the snow, or a scrap of hair. Even the bones are devoured.

Winter hunting is an ancient practice because often snow acts to a hunter's advantage. In czarist Russia, according to Soviet ecologist A. A. Nasimovich, winter hunting severely depleted populations of musk oxen and antelope and markedly reduced others. (And, Nasimovich points out, "capitalistic America [also] provides many instances of such irrational hunting.") Russian hunters formed ski associations and pursued prey through powdery snow until exhaustion delivered the animals directly to their knives. Rifles weren't even needed. Reports tell of two hunters from a mountain village who, on a single occasion, took over two hundred red deer, animals closely related to American elk. Another village of just fifty households boasted of killing fifteen hundred roe deer in a few days by hunting on crusted snow —thirty deer per household. The villagers also hunted moose in the snow. Trying to escape the ski hunters scores of them would flounder "from spruce tree to spruce tree" seeking snow capable of supporting them. Crust tore their legs as they repeatedly broke through, and the enormous effort of running so overheated them that they gulped mouthfuls of snow to quench their raging thirst. Under such conditions hunting wasn't a matter of luck. It was a guaranteed means of putting meat on the table.

Snow hunting from horseback also was common in Siberia and to some extent still is common. Horses run well in fluffy snow. Moose, red deer, roe deer, saigas, marals, and other hoofed animals do not. Laws now prohibit hunts in deep powder or crusted snow, and pits dug into snow as deadfall traps also are prohibited. Driving animals in snow of moderate depth and stalking or ambushing them on their daily routes are permitted under careful regulation. Tracking is easy in snow, especially on cold, windy days when willows are clacking together, covering the sound of a hunter's approach, and when the snow surface is dry enough not to squeak underfoot. Deer or moose hoofprints will have about the same hardness as the surrounding snow for the first hour after they have been made. Then they begin to set up. First the upper edges get crusty and hard, next the bottoms of the prints. A hunter needs only to step into several hoofprints to gauge how old they are. Or he can drag his foot along while passing on a snowmobile or dogsled. If the track is fresh he won't even feel it. If it is from the night before, he will notice a slight bumping as his toe drags from print to print. If it is a full day or two old each print will be hard-crusted, and the slight bumping will become a real jarring.

Snow less than ten inches deep can spell short-term disaster for some members of the antelope family even without hunters. If such snow is crusted and stays that way for two weeks or more, saigas are doomed. These are the small cinnamon-buff antelope of the treeless plains of southern Asia, their number so vast that Count Lev Nikolevich Tolstoi, hunting along the left bank of the Ural River in 1906, wrote ". . . the more I look into the steppe, the more of them I discover on the horizon." Saigas' pattern is to shift southward with the first snow, then north again as it thaws, perhaps moving back and forth three or four times in a winter. Heavy die-offs are common. As high as 40 percent of a herd has been found dead after only two days with snow about thirty inches deep. Similar accounts tell of wholesale deaths among red deer in the highlands of Scotland, reindeer in Norway, Dall sheep and moose in Alaska, and blacktail deer from Alaska to Arkansas. Hoofed animals simply are not well suited to deep snow. They put up with it rather than utilize it or derive any benefit from it, as is true of some other types of animals. If snow piles deep, ungulates lose mobility; the best they can do is move before that happens. Trying to get whatever food is available from under snow carries a high cost in energy, although some of it seemingly is squandered needlessly. An observer watching domestic reindeer in Siberia counted five times within an hour that a certain cow was chased from the grazing spot she had just pawed into the snow, and during the same brief period she also twice stopped of her own accord to chase nearby deer.

Plot twenty or thirty years' worth of snow-depth data for a region together with the census figures of its ungulate populations, and the two lines will have roughly the same

highs and lows offset from one another. The animals' lows
coincide with the snow-depth highs, and it takes a year or
two for the population to build back after the factors causing
depletion have ceased. R. Yorke Edwards, a Canadian mammalogist, used this method to check on what might have
caused die-offs of British Columbia deer reported in the
1920s. Reasons given in reports attributed the declines to
excessive market hunting, marauding railroad gangs who
were laying track at the time, encroaching civilization, and
Indians newly equipped with repeating rifles. Coyotes,
cougars, and wolves also were blamed. So was a massive
infestation of ticks. Yet Edwards found that die-offs from
these supposed causes coincided with the years of deep
snow. The other factors, from hunting to ticks, may have
intensified the deaths but snow was the real cause.

How readily an animal moves in snow depends on the
mechanical state of the snow and the structure of the animals' legs. Most ungulates walk on their third and fourth
toes, which form a hoof. The smaller, raised second and fifth
toes—dewclaws—also give some support in loose ground,
mud, and snow, and they furnish excellent brakes since they
jut out at an angle from the hoof. A few ungulates can
spread their hooves, separating the two principal toes to
widen the foot. Also, in some, the two main toes are quite
long and, on a soft surface, they bear weight for their entire
length, rather than just the hoofed tips bearing the weight.

Despite all such increases in hoof size, however, ungulates
are heavy for the bearing surface of their hooves; they sink
into soft snow and break through crusts. Their weight per
square inch varies by species from roughly a pound and a
half to sixteen pounds. A skier, by comparison, places only
from two and one-half ounces to slightly over half a pound
of weight onto his skis per square inch of surface, the
amount of course depending on the person's weight and the
type of skis worn. Furthermore, the weight a deer's hoof must
support depends not only on basic body size but also on
whether antlers are present, and whether the animal is
standing quietly on all four hooves, walking or trotting with
two feet down at a time, or leaping with four hooves at a

time. Sinking into snow beyond the upper part of the hoof greatly impedes any ungulate, and sinking belly deep immobilizes most. Thus, the length of the legs largely determines ability to get around in snow, and species ill equipped in this regard adapt to snowy country best by leaving it when winter starts. Generally this means heading to a lower elevation or moving southward, although in a few regions specific local snow conditions may warrant moving north, east, or west, and sometimes even moving upward in elevation.

For example, as winter approaches in the Ural Mountains moose and roe deer that have been summering on west slopes start to shift eastward, to the snow-shadow side of the range. One of Yellowstone's elk herds moves to high slopes blown free of snow by howling winds, and a British Columbia caribou herd in mountainous Wells Gray Provincial Park makes a double migration. These caribou leave the heights, return to them, then leave again and return again—all within a single year. Theirs is an adaptation suited to specific local snow conditions. Six to ten feet of snow blanket the timberline slopes of Wells Gray, while valley bottoms lie covered with only two or three feet. By November, as the high-country snow accumulation is well underway, the caribou move into lowland forests, but by February they are again high. Furthermore they choose the subalpine slopes with the deepest snows.

The reason for this behavior lies in the interrelationship between food habits and snow. Studying these caribou, Edwards found that in summer they graze high mountain meadows, and as autumn's snow begins to cover the ground they paw through it to reach their preferred plant species. As the snow deepens they drift gradually downslope, still pawing to feed on ground vegetation. On the valley floor they graze beneath trees where dense intercepting foliage minimizes snow depth. To this point these caribou behave about the same as other caribou. Then they return to the snow-gripped mountains, doing so to escape the increased depth and density of the forest snowpack. As it builds, caribou have more and more trouble pawing through it to feed, and they switch to browsing lichens from trees instead. Many of these grow low enough to reach easily; others are brought to the forest floor by wind and falling trees. By January a thaw almost always settles the snow enough that the caribou can walk on top of it. Able to move about freely, they climb back to timberline. Arboreal lichens there are more abundant than in the lowland forest; and, furthermore, each successive mountain storm deepens the snow and wind-packs it, thereby lifting the caribou higher and higher into the "lichen pastures" of the trees.

By April, spring begins in the valleys and before its warmth softens the lowland snow, the herd starts down again. Their second arrival in the valley bottom usually comes just as ground vegetation is melting through the snow, and the first crows, robins, and flickers return at that same time. From April until June the Wells Gray caribou graze the rich bottomlands and the forest openings where snow melts rapidly. As the melt moves up the mountain slopes, they follow it to their summer range, and the double migration is over for the year.

In the mountains caribou live in relatively small, scattered bands and herds. Most of the species, however, live farther north where the trees of the taiga edge vast windswept barrens. These northern caribou—and comparable reindeer in the Old World—are the ones renowned for long migrations between summering grounds and wintering grounds. Herds even today are great enough to stretch from horizon to horizon. In recent years 14,000 caribou were counted on a single day crossing Lake Athabasca, which straddles the northern border between Alberta and Saskatchewan, Canada. Individual herd counts are estimated as high as 150,000 and in North America as a whole there was a total of 1,000,000 caribou until their numbers declined in the 1960s and 1970s, largely because of overhunting. There are ten major herds, six in Canada and four in Alaska, plus un-counted small and scattered herds such as the one in Wells Gray Park. Wild reindeer still exist in Eurasia and have been reintroduced into Greenland, but wild and domestic herds intermingle so extensively that assessments of the old-world wild reindeer population and range are difficult.

When undisturbed, caribou make spectacularly long migrations that are unmatched for distance by any other land mammal. The Barren Grounds herd of northern Canada treks from summer grazing along the shores of the Arctic Ocean to wintering grounds in the muskeg and sparse forest of the subarctic 700 to 800 miles away. Caribou that spend the summer on arctic islands travel as much as 150 to 200 miles over sea ice as winter approaches, and in one reported instance a herd traveled a full 400 miles from Novaya Zemlya to Spitsbergen. How the animals navigate while on these journeys isn't known, but, if need be, they can travel in fog for days, crossing featureless terrain yet following traditional routes. These often lead along moraines or eskers left by Pleistocene glaciers. These are ridges of rock rubble that snake for miles across the country and, being raised, usually are blown at least partly free of snow. Walking there is easier for caribou than on the flats, and they can paw to food they smell beneath the snow.

The great herds leave the northern tundra not because of cold but because of snow. Ranges that easily support hundreds or thousands of caribou in summer may carry only a quarter that number through the winter, or none. Too little food is available while snow blankets the land. Migration southward begins as early as August in some regions, and by September or October it is underway throughout the north. Snow herds the animals along. Too thick, too hard, too dense—and they go elsewhere. Caribou tolerate a depth of up to only about two feet, according to current studies by William O. Pruitt, outstanding Canadian authority on the ecology of snow. Similarly, they have a critical tolerance for surface hardness. In forested areas snow must be hard enough to support three quarters of a pound per square inch, and frozen lake surfaces must support ten times that weight before caribou will venture onto it. Somehow they know.

For density, their requirement is 0.19 to 0.20 for forest snow and 0.25 to 0.30 for lake snow. (By comparison, 0.9 is the density of ice.) Nasimovich, observing domestic reindeer on Novaya Zemlya in the 1930s, noticed that they would paw through as much as twenty inches of snow providing its density wasn't above 0.20. If the density rose to around 0.40, however, feeding efforts were defeated by a mere eight inches of snow cover. Nasimovich also found, as Pruitt has, that herds move when their preferred snow conditions aren't present. Wild reindeer on the Kola Peninsula and in the southern Ural Mountains migrate back to upper elevations when metamorphosis at the bottom layers of lowland snowfields has reached the "sugar" stage. This is a loose granular structure that flows like sugar from a cup whenever a deer paws through the layer to reach food. The walls of feeding craters collapse, and sugar-snow floods over exposed vegetation, thwarting all efforts to feed. Continually brushing it aside is futile. Rather than struggle against it, reindeer return to the mountains where microtopography assures a wide variety of snow conditions. Windblown hillocks seldom have much snow. To stay with the "right" snow, caribou may travel from twelve to forty miles a day for a week or two at a time. They are fenced by snow as effectively as if by the pasture fences of a rancher.

As caribou walk, a distinct clicking sound comes from their hooves. It can be heard from a mile or two away on still days. Another characteristic is that the hooves splay out to about the size of a man's hand with fingers spread. This is ideal for walking in snow and digging for food. It also helps to make caribou the best swimmers in the deer family, capable of swimming two miles in an hour if need be and of keeping it up for several hours at a time. The bottom surface of the hoof is slightly dished with a sharp outer edge that provides extra traction for walking and extra clout for breaking through light crust to graze. Special tufts of oily nonstick hair prevent snow or ice from clogging the hoof. Thus equipped, the great herds accomplish their treks, nibbling lichen or sedge or biting from a low huckleberry bush or willow as they go. Fly over caribou country and you see their feeding craters pocking the snow everywhere. Because of the craters the Micmac Indians of eastern Canada called the animals *xalibu*, "the one who paws," or the

Caribou make spectacularly long migrations, unmatched by any other
land mammal. Canada's Barren Ground herd grazes the shores of the
Arctic Ocean in summer, retreats 700 to 800 miles into the muskegs
and taiga of the subarctic in winter. They can smell food through the
snow and paw down to it. North American herds still number in the
hundreds of thousands, but have declined in recent years owing to
excessive hunting by snowmobile. A man now can travel a hundred
miles, make his kill, and return home in the same day.
(D. Thomas, Canadian Wildlife Service)

Deep snow presents great difficulty for hoofed animals such as Yellowstone's elk. The animals' weight per square inch of foot size causes them to sink into soft snow and to break through snow crusts. Dropping belly deep causes immobility—and therefore opportunity for predators. One Yellowstone elk herd migrates to higher, windswept elevations to avoid deep snow.
(*F. Jay Haynes, the Haynes Foundation*)

Bison swing their heads from side to side to brush away snow
and reach down to winter grazing. In Yellowstone National Park
they often stay in geyser basins, where geothermally heated ground
minimizes snow accumulation. During blizzards they face
into the wind, protected by the shaggy manes
of their heads and shoulders. *(Ruth Kirk)*

Supercooled water droplets freeze on contact with snow crystals,
producing rime as shown in this photomicrograph *(above)*.
Depth hoar forms at the base of snowpacks as water vapor migrates
upward *(below)*. The hoar lessens the mechanical strength
of the snow and increases avalanche danger. It also eases
the movement of small mammals living under the snow.
(Edward LaChapelle)

Compacted ski tracks resist wind erosion and remain raised
above the overall snow surface. The weight of oversnow vehicles
compacts snow so greatly that it poses potential problems
for plant and animal life beneath the snow, which is why
many areas limit winter travel to established roadways.
Compaction lessens the insulating quality of snow. *(Ruth Kirk)*

Kobuk Eskimo women snare winter ptarmigan by bending willow shoots
and attaching nooses, which the birds walk into while feeding.
Ptarmigan plumage provides year-round camouflage:
a mottled brown in summer, snowy white in winter.
(Above: Ruth Kirk; below Frank Brockman, National Park Service)

As morning sun warms the floor of Yosemite Valley, California,
snow unloads from the branches of a pine tree *(above)*.
In places, forests are managed specifically for their effect on
snowpack, which is a natural reservoir that today's humans
depend on. Trees affect the amount of snow reaching the ground
and also the rate of melt on the ground *(below)*. *(Ruth Kirk)*

Avalanche lilies bloom before winter snows have left subalpine
meadows. Stored carbohydrates permit active life as light first
penetrates thinning snowpacks. Growth tips that burst from the bulb
the previous fall lengthen and literally melt up through the snow,
warmed by their own metabolism and by the raised temperature
within their meltholes. White petals unfurl above white snow.
(Ruth Kirk)

shoveler, and from this word came the name caribou, passed along in early French accounts.

Smell guides the discovery of food beneath the snow. In Alaska the late, deeply beloved American naturalist, Olaus Murie, once watched a blind yearling feed successfully as part of a herd, and subsequent stomach analysis showed no difference between this animal's choice of food and that of its sighted associates. This was true both for kinds of plants represented and proportions of the various kinds. Lichens were the favorite food. Vapor rising from the relatively warm bottom of a snowpack to the air-chilled surface carries the odor of plants with it, especially of those that aren't frozen. Guided by odors, caribou dig only where food actually is available. They don't squander energy in unsuccessful attempts at finding it. By pawing thirty or so craters, caribou get five or six pounds of lichens and other food, ample for a day. They eat on the move, pulling a mouthful or two and walking on to sniff the snow and paw a new crater. In this way dormant plants aren't grazed beyond the point of recovery. If snow is deep, however, caribou may remain almost sedentary.

Ironically, despite such a long and flawless balance between food supply and need, caribou and reindeer now are suffering because of their feeding habits, particularly their fondness for lichens. Unlike herbaceous plants, lichens draw sustenance directly from the aerial fallout of dust and organic material. In today's world this means they are peculiarly susceptible to radioactive contamination—and therefore so are the herds that feed on the lichens. The Lapps, Siberians, Eskimos, and the northern Indians who eat caribou meat may also be affected. Swedish studies show a concentration of cesium 137 in reindeer meat nearly three hundred times greater than that in beef raised in the same general area. In Norway and Alaska strontium 90 levels four or five times above that in beef, cattle, and sheep have been found in reindeer meat.

A. N. Formozov, the Russian ecologist who pioneered study of how snow affects life, has classified mammals as chionophobes, chionophiles, or chioneuphores, using the Greek word for snow *chion*. The -phobes are those unable to adjust to snow: most cats, for instance, and opossums. The -philes are those few with specific adaptations to snow such as caribou and snowshoe hares. The -phores are those that can survive in snow but lack delicately tuned adjustments of either physiology or behavior.

Elk and moose are chioneuphores. Their long legs help them move about in snow. They also give them a chance to outrun predators and to reach the twigs they favor for browse, which probably means that the length didn't develop specifically in response to snow. Nonetheless long legs act as stilts holding the possessor's body above the snow surface, and moose are probably the next best snow travelers after caribou. Under some conditions they too do everything possible to avoid the "wrong" kind of snow. When it is deep they may stubbornly follow a packed sled trail, even refusing to give way to an oncoming dog team, which guarantees pandemonium. Around two and a half to three feet seems to be the maximum snow depth moose will tolerate. If there is more than that their bellies drag. To avoid deep snow many moose make seasonal shifts in altitude or latitude. For instance, an altitude drop of 4,000 feet may require a British Columbia moose to travel 30 or 40 miles. To gain the same effect by shifting to a more southerly latitude it would have to travel for 350 miles.

Another means of avoiding excessive snow depth is to "yard up." Moose need an enormous quantity of food to survive; as Alaskan Bob Uhl puts it, they have to get through the winter with nothing more than frozen willow tips, just like ptarmigan. Picking a place with ample forage and staying there helps. Otherwise the energy cost of fighting "wrong" snow may be greater than the gain of whatever food is found. Trampling packs the snow of the yard. An ideal location is a grove of willow or aspen sheltered from midday sun so that the snow surface won't melt in the morning and refreeze into a hard crust by afternoon. Eight or ten moose typically yard together in such a spot, staying within a fifteen- to twenty-five-acre area. Snow falling from the tree branches overhead contributes erratic compaction to the white blanket on the ground, and new snow sifting from the clouds bends branch tips into convenient reach for nibbling. Or, that failing, moose will rear up and pull down what they need with their forelegs. How well the animals fare depends on how long deep snow confines them to their yards and how much forage is available there. Russian studies show that moose eat back the twigs of a single bush or tree five times as intensively when the snow depth is around two or three feet as when it is only four to five inches deep and moving to other browse is easy.

In winters of deep and frequent snowfall moose may yard in spruce and fir thickets rather than in stands of birch, aspen, or willow which provide more nutritious browse. Or, they may migrate through coniferous forest in preference to broadleaf forest. This is placing a greater survival importance on minimum snow than on optimum forage. A recent British Columbia study indicated only about one-third the amount of palatable browse available within a mature spruce-fir forest compared to that within an aspen grove.

Another means of getting around in snow—other than with the help of built-in stilts or snowshoes—is the battering ram approach. This is the method used by bison. Their legs aren't long; their hooves are small and sharp. But their strength is prodigious. If snow isn't crusted, bison can bulldoze through drifts three or four feet deep. You see them plodding with belly submerged, swinging their great shaggy heads from side to side to plow a trench into the snow. Cows and young string out behind the most massive bulls of the herd, following closely in their footsteps. This is an advantage that herd animals have over solitary animals such as moose. Within a herd the lead position can be rotated as the exertion of breaking trail takes its toll, so no individual has an undue burden, and all benefit. Herd members follow one another so exactly that it can be difficult to tell whether a lone animal or a considerable number have passed. The pressure of each footfall melts the snow within that print and, as each refreezes, the chain of tracks furnishes a firm if perforated path. Outside the tracks themselves the snow stays soft.

In Yellowstone bison and other animals have an additional —unusual—means of coping with snow. Some of them spend the winter in hot-spring basins where geothermally heated earth minimizes snow. Several even stay in Pelican Valley where snow gets deeper than in adjoining areas, but where hot springs assure some snow-free ground for grazing. This gives a survival margin that outweighs the disadvantage of deep snow in most of the valley. Mountain sheep in the Geisernaya Valley of Kamchatka also retreat to geothermally heated ground where early snowmelt fosters the growth of spring herbage.

Such opportunity, of course, is rare, and even among Asian sheep and American bison it is a minor adaptation. More common for both is to avoid snow by seeking exposed ridges that are swept free by wind. Mountain sheep, mountain goats, mountain antelope, chamois, and turs are well equipped with strong legs and hooves that give them unexcelled footing on steep slopes, or on loose surfaces such as rock scree and snow. They are inept in deep or crusted snow but are eminently well adapted to staying mobile and adequately fed among the high cliffs. There is no competition there from other grazers and browsers.

Musk oxen also depend on vegetation blown free of snow, so much so that they will crop an exposed slope until little food is left rather than feed in lush vegetation only a few yards away but lightly covered with snow. Dwarf willow and grass are favored. These are plants ignored by caribou, so the two large herbivores can live on the same range without competing. If proper grazing isn't to be found, or the energy needed to find it would exceed that likely to be gained from it, musk oxen have no real problem. They live off fat reserves built by around-the-clock feeding during the fleeting weeks of summer largesse. At that time they nibble practically nonstop, pausing only to chew their cuds and to cool off by resting on snow patches. Winter snow poses a problem, however. Musk oxen have no way to deal with deep drifts, and at times they have died by the hundreds and possibly thousands because of such conditions.

Their legs are short for their bodies. If a raging storm swirls around them and builds deep drifts of snow all these animals can do is to form a defensive knot with the young at its center where the warmth of the herd will provide maximum protection. Often this tight huddle is wedge shaped with two or three aged bulls at the apex, their high humps acting as windbreak for the cows and calves. All the animals face into the blizzard to minimize disturbance of their shaggy coats. If necessary they stand for days waiting out the storm without even lying down.

Musk oxen are the only land mammal that can bear the full blast of a far-north storm by doing nothing beyond standing and waiting for it to end. Other northern mammals, such as sled dogs, arctic foxes, and polar bears at least curl into the snow for protection if no better shelter is possible. But musk oxen survive with no shelter or fodder. Their wondrously warm coats are the key. An underlayer of fine wool covers their backs and hangs practically to the ground. It is thick and felted and serves as an airtight fringed rug of the best quality. Over this soft layer is a dark brown fleece of long guard hairs, slender at the roots and swelling toward their tips. They lie as a close-fitted armor at the surface, holding air along their shafts and also protecting that within the underfur. Ancestors of today's musk oxen roamed arctic and subarctic tundra and steppes one million years ago, contemporaries of woolly rhinoceroses and mammoths. They now survive in Greenland and arctic Canada and have been reintroduced successfully into Alaska, Spitsbergen, and central Norway. Protected by law throughout their range, they are holding their own and even increasing in number. Their scientific name, *Ovibos moschatus*, means sheep-cow, but musk oxen are neither sheep, cow, nor ox. They are ancient holdovers with takin, a goat-antelope of the central Asian highlands, as their nearest relative.

Beyond question man is the animal with the greatest impact on other life-forms in the North, but next in importance

is the wolf. Snow gives both certain advantages. A wolf's
track load is only a little over half that of a caribou and one-
fifth that of a moose. Wolves hunt when snow crust supports
their weight but not their prey's. If such crusts are too few
during a winter wolves may starve. In fluffy snow, wolves
also have something of an advantage over their prey because
escaping animals must break trail for themselves, yet they
leave a packed track behind them for their pursuers. Some
field researchers feel that this packing of snow, especially by
herd animals, is extremely important for wolves even when
not in hot pursuit. Without such highways they would have
trouble staying mobile through the winter.

The role snow plays in the relationship of wolves to prey
has been meticulously documented through nearly three
decades of observation at Isle Royale, Michigan. The island
is a 210-square-mile dot 15 miles off the Ontario shore of
Lake Superior. Early in this century moose swam to the
island and found Eden. There were no predators and no
competitors. In a decade or two their numbers grew to a
higher population density than is known for moose anywhere else in the world. There were perhaps ten or more
per square mile. A biological slum quickly resulted. Plants
were browsed beyond recovery. New stands of poplar
needed by beavers couldn't get started. Cover for snowshoe
hares was damaged. Coyotes and foxes fed on the carcasses
of moose that had starved to death, but they couldn't bring
the population into control. Then a die-off set in, and the
moose dropped from over 2,000 to only a few hundred by
the mid-1930s. The most vigorous animals from the former
excess population were left, and they quickly bred the herd
back up beyond what the range could support.

In winter sometime between 1948 and 1950 wolves crossed
to the island while the lake was frozen. By the late 1950s
coyotes had vanished, probably because of the newcomer
wolves since the two occupy much the same niche. Biologists
were delighted with the whole wildlife picture. With only
one major predator present—the wolves—and one major prey
species—the moose—Isle Royale became an ideal outdoor
laboratory. As it is physically isolated, no external influences

were likely to intrude, and, without diversity of species, no welter of confusing interrelationships would complicate study. Predator-prey, wolf-moose, relations would be clearcut. Durwood Allen, Purdue University mammalogist, got observation underway in 1958 and assisted by graduate students, such as David Mech and Philip Shelton, he has continued study ever since.

An outstanding result has been evidence of population stability on Isle Royale with about twenty-four wolves to one thousand moose and two thousand beaver, their principal prey. Snow plays a definite role in the balance. The first three winters of the Purdue study were years of below-average snow depth, and moose were often found in the relative openness of a burned area where brush was abundant. There wolves made kills. The following years had more normal snow depths of between twenty-five and thirty inches. This depth caused moose to seek the shelter of trees rather than wallow through the deep snow of open areas. The burn was abandoned. In 1968 little snow fell, then each of the following four years brought extremely heavy snow, and Allen and his students found that the rules between wolves and moose had changed. During the years of deepest snow there were more wolf kills than before, over half of them calves. This was double the calf predation rate of years with normal snow depth. Apparently in winters of deep snow, moose concentrated along the lakeshore where footing is easier. Browse is scarce there, however, and mostly out of reach for half-grown calves. Examination of bone marrow in dead calves showed little fat, a lack that indicates malnutrition and probable exhaustion. The weak fell easy prey to wolves, also concentrated along the shore where shelf ice facilitates travel for them just as for moose.

Curiously, the killing of young moose continued each winter of deep snow. This was true not just of new calves and yearlings but of moose up to five or six years old. During earlier observations, this age group had seemed exempt when snow was normal, but in 1969 and the three following years they fell frequently to wolves. It now seems probable that deep snow during the first winter before calves are

born, or the first one after their birth, has far-reaching
effects. It somehow causes a physical impairment that lin-
gers and renders moose vulnerable to wolves.

Men have feared, hated, and persecuted wolves only in
recent centuries. Indian and Eskimo peoples don't feel this
way, although they have lived with wolves long and closely.
It is man as herdsman, not as trapper or hunter, who detests
them. In western United States and Canada there still are
homesteaders who speak of "that awful cry in the night"
when wolves were attacking the cows. Both wolves and
herdsmen value livestock; they are in competition over them,
for of course wolves make the easiest kills they can. Their
motive is survival.

In nature, wolves don't decimate the wild herds they depend upon, as the Isle Royale data substantiate. Observe
a hunt, and you see a pack give chase for perhaps an hour;
then if there are no stragglers they turn to other pursuits.
If necessary they can go for days without feeding. When
they do make a kill nothing is wasted; wolves even lick
blood from the snow while they circle a wounded animal.
Parts of a carcass they don't immediately consume often are
covered with snow to be fed upon later. Sometimes they
make these caches under their beds, so as to conceal them
from ravens, crows, jays, mice, weasels, martens, fishers,
wolverines, and foxes, scavengers that raid wolf kills whenever they can. Ravens and arctic foxes even follow wolf
packs to share their kills. Ravens don't fly at night when
wolves most commonly are on the move, but they track
them and catch up as daylight returns.

Once wolves roamed all of the northern hemisphere except for the densest rain forests and driest deserts. Today
their numbers and range are mere fractions of what they
were. Wolves are practically eliminated from the United
States, except for Alaska, and there and in Canada their
fate alternates between policies of protection and of bounty-
subsidized extermination. In the Soviet Union 42,000
wolves were killed in 1946; 8,800 in 1963.

As a rule hunting large animals such as moose or caribou

gives wolves the greatest return for energy spent, but at times they also take small animals. They may hunt beavers, as at Isle Royale, or lemmings, voles, hares, and others. In summer ground-nesting birds' eggs and chicks furnish occasional fare. For the most part, however, it is small-sized predators that make use of this small prey. They depend on it. In years of lemming decline, foxes kill most of their kits; somehow they know the odds are against survival, and they concentrate their full efforts on raising just one or two young. At such times ermine, which feed almost exclusively on lemmings, may not even bother to breed. Skuas arriving on migration soon leave if their lemming food supply is at low ebb. Snowy owls will feed only one hatchling in years of lemming scarcity—and these will be the years that people as far south as Texas phone to Audubon societies and state game departments to ask about the large white bird they have just noticed sitting on a fence post.

The lemmings themselves follow a built-in cycle which, about every four years, prompts what ends as a kamikaze stampede. Biologists don't fully understand the underlying reasons. Years of high lemming population are not widely synchronous from region to region, and there is no similarity in how many animals are involved from one cycle to the next, or over any very great distance. Migrations seem to occur only in Eurasia, not in North America, and they are not deliberate suicide marches to the sea. Rather they are great nervous millings about. If the most natural route of travel funnels seaward, the lemming hordes flow that way and, ultimately, with no more land ahead of them they continue into the water and swim as long as they can, then die. Many small herbivores with short life-spans experience comparable population excesses alternated with mass die-offs. They are cyclic among hares, muskrats, marmots, voles, and ptarmigan, as well as lemmings, although only the lemmings migrate.

If food is abundant, reproduction takes place at an extraordinary rate; if not, it doesn't. A lemming litter born under the snow in March can produce its own progeny before the spring is out. But if the snow cover is thin while winter's cold still prevails, female ovulation will cease, and there will be no births. If snow lasts late and spring vegetation is meager, reproduction also will decline. There may already be a surplus population, however, product of a previous fecund year; and, if so, the famous migration to oblivion may start even before food has given out. Lemmings basically are not sociable, and when crowding brings them together the resulting stress sets up a hormone imbalance that causes some to move out. Jammed together by vagaries of topography they run with insane feistiness, chattering their teeth, snapping at one another, and panicking at any unexpected situation.

So-called lemmings belong to three distinct genera: *Lemmus*, *Dicrostonyx*, and *Synaptomys*. Most weigh two or three ounces and have been described as looking rather like "prosperous mice." Similar in habit and sharing the same range are voles, slightly smaller than lemmings. Common ones belong to two genera, *Microtus* and *Clethrionomys*. Neither lemmings nor voles hibernate or have sufficient reserves of fat to survive winter without its blanket of snow. Even beneath it they must scuttle about continually to feed on whatever roots and stems and carrion they can find. This seldom is any problem. Most tundra plants carry growth buds through winter just above the ground surface, safe from freezing because of snow's insulation but readily available to hungry lemmings and voles. Also, because these plants must sprout, grow, and reproduce with great haste as soon as summer begins, they maintain a dense mass of roots and rhizomes rich with stored carbohydrates.

The animals' need to feed actively beneath snow's protection works out well, as is true of all of nature's systems. Specially adapted to the task of finding winter food, the claws of lemmings and northern voles elongate and divide into prongs as autumn days shorten and turn cold, thus providing built-in snow shovels. Species that live in regions of particularly dense snow have the most pronounced changes in the structure of their feet. In a similarly adaptive way, one species that lives where snow is shallow turns white in winter—the only rodent known to do so. This is the varying lemming. With a comparatively light snow cover, it is apt to venture to the surface in winter and therefore needs the protection of camouflage.

Most lemmings and voles, and other animals equally small, depend on the insulation of at least three feet of snow through the winter to assure adequate warmth for their chambers when the temperature drops to its lowest. Even as little as six inches of snow offers some advantage, however. When snow reaches that depth the soil temperature stops following the swings of air-temperature highs and lows, and lemmings, voles, and shrews start tunneling within its white blanket. Runways still are likely to collapse, but some insulation is gained and a roof overhead, no matter how fragile, offers concealment from foxes and owls. When snow has thickened to two or three feet, the ground temperature will stay up to seventy degrees Fahrenheit above air temperature. Any small creature unlucky enough to have burrowed where snow blows away is doomed to freeze to death. Adequate snow cover is absolutely essential to winter survival. For rodents in winter, body mass producing heat isn't great enough by itself to keep up with body surface giving off heat. Without snow's insulation, no amount of eating could produce a metabolism high enough to offset chilling.

In studies on the Kobuk River, Pruitt found that whereas the air temperature fluctuated from about 7° F. to −25° during one nine-day period, the forest floor beneath two and a half feet of snow held a constant temperature. Variation was from place to place rather than from day to day. For instance when the air was −24° F., a thermometer placed in a snow saucer at the base of a spruce registered only −2° at a depth of one inch into the soil. Barely beyond the edge of the saucer, where snow was deeper, the comparable temperature was +54° F. As would be expected, the movements of small animals beneath the snow reflect these temperature differences, and differences in snow conditions, just as markedly as do the migration times and routes of caribou.

The subnivean realm is cold and dark, if not intensely cold. One foot of snow passes only about 8 percent of the incoming light, depending on density. Two feet of snow cut the amount in half. No sound carries except the occasional tinkle of depth-hoar crystals shattered as a lemming or vole or shrew ambles along a thoroughly metamorphosed layer of snow at the bottom of a pack, or as loads falling from overhead boughs thud down into the snow. Life follows an easy rhythm of sleeping, feeding from a supply of cached food, and tunneling through the snow to a new source of food. Voles sometimes dig runways up to the branches of bushes to gnaw bark, leaf buds, and catkins—plant foods they feed on only while under the snow. They also tunnel up to deposits of birch seeds. These mature in January and rain from the trees onto the snow surface, where redpolls and other birds feed on them. When a new snowfall mantles the land, whatever seeds remain are automatically stored within the territories of voles. Ultimately, the rich odor of the seeds betrays their presence.

If winter cold comes suddenly and early, abundant nourishment is available under the snow. If, on the other hand, it swings between freezing and thawing, there may be problems. Only abrupt and steady cold quick-freezes plants, retaining nutrients the same way as in man's frozen-food industry. A tug-of-war between autumn and winter, particularly if there are rains or wet snowfalls, destroys nutrients. If that happens not even the insulation of a deep snow cover can tide small animals through the winter.

Bacterial decay usually continues beneath the snow because of mild, stable temperatures, and carbon dioxide is a by-product. At first it is carried up by rising vapor but as metamorphism continues within the snowpack, ice layers sometimes cut off the escape of gases. Carbon dioxide then, being heavy, settles into hollows and low spots beneath the snow. Voles can't escape the poison by tunneling sideways, partly because the condition usually extends over a large area and partly because they would run into their neighbors' defended territories. The only way to go is up, and they dig to the surface. Overnight the snow becomes pocked with ventilation shafts. Wisps of air rise from them. Foxes scent the presence of warm life beneath the snow and make kills to sustain their own lives. Voles seldom venture up the shafts; they maintain them only as vents for the carbon dioxide.

Foxes hunt also by zigzagging over the snow surface listening for faint squeals and scratchings, which they can hear through ten or twelve inches of snow. When a signal is picked up, the fox leaps high into the air to land stiff-legged with nose and forepaws held together. The sudden pressure breaks the snow crust, and the fox's jaws instantly close on its prey. Deep or compact snow thwarts such hunting.

Weasels and their European counterpart, stoats, hunt much the same prey that foxes do, but they go about it differently. They are small and supple enough to follow lemmings and voles into their burrows. They essentially swim through snow to initiate their chases, diving in and staying submerged until they succeed. None of the fox's energy expenditure of listening and pouncing, failing, and listening and pouncing again is required of the weasel family. They are at ease in the subnivean realm of their prey.

Lynx are another predator of the snow, the only American cat well adapted to the white world of winter. They are small cats despite the apparent larger size their fluffy coats give them. Broad and densely furred paws combine with fairly long legs to help lynx get around well in snow. Snowshoe hares are their main fare. They hunt by alternately leaping, then freezing in their tracks and crouching stationary and silent. Their sudden motion startles hares into bolting for safety, and their equally sudden disappearing act as they freeze the motion lures hares into hopping closer to investigate. It is, of course, their last hop. For lynx the ruse is necessary. They flounder if they try active pursuit of hares, but their bound-and-stop techniques work well, and they also sometimes take up vigil at the base of a tree. They wait for hours, then pounce when a hare happens by.

Wolverines often victimize lynx. When snow is deep they trail them, then steal kills. Lynx are afraid of wolverines and retreat as soon as they see one. Wolverines also raid carcasses cached by wolves. In fact trappers often make their sets by such caches—and occasionally succeed in taking a wolverine. Wolverines are "devils," according to

Gunther Lishy, though he once trapped sixteen in a single winter. "You can do all your tricks and still a wolverine will outwit you." The battle, with all its scheming and maneuvering, is an ancient one. Man, in his long relation with snow and the life on it and in it, traditionally has understood the links between one species and another, and one snow condition and another. He himself has belonged to the web, and today's pioneering studies of snow ecology are only explaining what trappers and hunters have sensed empirically: that snow for many species is the key element in survival.

CHAPTER

9

SLED DOGS
AND REINDEER

FLY INTO ANY ESKIMO VILLAGE IN WINTER, AND YOU ARE LIKELY to find human villagers congregated in the warmth of church or school or home, or running errands on their snowmobiles or by taxi if there are roads. The village dogs, however, will be out in the snow, chained just out of reach of one another. If they are to be run, or if someone's passing excites them, they will be leaping and howling, their din filling the air with wild crescendos. If the day is cold, or at night, or if snow is falling or blowing, they will be lying curled with their tails wrapped around legs and noses. In this compact position dogs reduce body surface and lose less heat; also, their thickly furred tails protect their lightly furred lower legs and feet. No matter how cold or stormy they can survive outside without shelter.

Under similar circumstances wolves and polar bears and arctic foxes tunnel into snow banks. Dogs usually can't dig in. Their chains, only six to eight feet long, hold them short-tethered to stakes and their own constant presence packs the snow around the stakes to ice. Digging is impossible, and it's not needed. When storms come, dogs simply let themselves get buried by snow, thereby effortlessly achieving the insulation that other animals gain by tunneling. Even with snow lying directly against them, their fur doesn't get wet. It holds in body heat so remarkably that not enough escapes to melt the snow. This is essential, for wet fur is a poor insulator, and if it were to get wet, then freeze, it would encase the dogs in ice. But this doesn't happen. Instead the dead-air spaces of the fluffy snow are added to those of the fluffy fur, wind is excluded, and the dogs' warmth is enhanced.

Even without the advantage of snow, northern dogs can withstand intense cold. In a test, four sled dogs were left in a cold chamber at −58° F. for three hours with no protection. They simply lay curled, tails over noses; yet at the end of the period their deep body temperatures showed no drop. Neither was there a significant lowering of skin temperature, nor of that just under the skin. Left to their own devices the dogs can manage. They don't need even a wall of snow to break the wind, nor do they need to be moved to the lee of their owner's house during a blizzard. Sometimes pups are brought inside, but once off the breast they probably never again will see the inside of a house. Eskimo dogs need no concessions. On the trail they sometimes must endure wind-chill temperatures of −100° to −125° F., yet keep pulling. Under such conditions a pampered dog would be useless and almost certainly its owner would have to shoot it.

In Siberia, Russian peasants living along the north Yenisei River formerly built large flat-topped ovens of stone between rooms of their houses, useful not only for baking and heating but also for sleeping on top of in winter. Yet whoever bedded down in this choice spot surrendered it to the dogs, which were peaceable Samoyeds, whenever they came back from winter sledging. In northern Canada the Kutchin Indians gave their teams far simpler protection, tying them in brushy areas that offered at least partial wind protection and sometimes also providing thick pads of dried grass for the dogs to lie on. Such attention was more readily feasible in the forested country of northern Russia and Canada than on the wind-swept coastal plain where most Eskimos lived and, where practical, it saved dog food. In intensely cold weather, without protection, dogs eat a great deal more than they otherwise have to.

How the relationship between man and dog first began is conjectural. The ancestor of the dog may never be known; it may have vanished as the ancestors of the modern horse and cow have vanished. Or it may be that dogs evolved from wolves, jackals, or the pariah dogs of the Orient. Living wild, these all nonetheless would welcome the chance to share human hunters' kills, and all also would be likely to scavenge scraps around camp. From man's standpoint, so long as his own food supply wasn't threatened, the improved sanitation resulting from such scavenging would be an asset, whether or not recognized as such, and the dogs would provide an alarm if they sensed intruders. Doubtless the first contact between man and canine brought some such direct benefit to both.

Wolves are the most likely lineal ancestor of dogs. Anatomical evidence favors them over jackals or pariah dogs. Their domestication seems to have been man's first, and from it developed a host of advantages aside from watchdog and sanitation services. Man soon must have seen the possibility of letting dogs locate and run down game, holding them at bay until he could arrive to make the kill, and he must have started deliberately training his pack in this. Probably, too, he learned quite early to breed for it. As herding became added to hunting, dogs learned to watch over flocks, and eventually to herd them.

Using dogs for haulage seems confined to the northern hemisphere, and especially to the arctic and subarctic regions, although Europeans until fairly recently hitched dogs to wheeled gigs and delivery vans, and Indian peoples of the North American plains used dog travois, which are long poles worn with a harness and dragged. Sledding, and the deep dependence of man on dogs, however, belongs exclusively to the Far North and to breeds within the Spitz family, mostly malemutes, huskies, and Samoyeds. The malemute, comparatively lanky and wolflike, developed in Alaska and northwest Canada; the stouter husky is the breed of the Chukchi people on the Siberian side of the Bering Sea; the white, chunky Samoyed originated in western Siberia, so gentle that it was given the oven-top sleeping place and even was used as a warm bedmate for human babies.

Sleds pulled by dogs varied depending on type of use and available building materials. Wood is scarce in most of the northern Arctic and Eskimos often substituted bone, whale baleen, or even frozen hides for it. One of their simplest methods was to soak a skin in water, fold it hairside out, press it flat, and let it freeze. Bearded-seal, polar-bear, musk-oxen, beaver, and caribou skins all were fashioned into such sleds, which differed from region to region. Generally old, worn hides or discarded tent skins, were used this way. They were make-do sleds, not usual ones but, with luck, could serve for a winter or even two if kept dry when not in use.

Caribou antler or small pieces of wood or whalebone, or even chunks of hard-frozen meat, were lashed onto the folded and frozen hides as crosspieces. Netsilik Eskimos, in northeastern Canada, sometimes strengthened their skin sleds by laying a line of fish onto a wet skin, overlapped like shingles. Then they folded the hide and tied it with sealskin thongs, curving up one end before letting it freeze. Interior Athapascan Indian people, in Alaska and Canada, sometimes simply hand-dragged a stiff-frozen moose hide or crudely lashed tree branches loaded with household gear or meat from a kill. They also occasionally laid buffalo ribs nearly parallel and tied the smooth ends together at the front with a cross brace. The opposite, knobby ends of the bones were held by a thong.

When possible, sled builders used wood. To get it many Eskimos in the central Arctic traveled great distances inland to forest lands. Others relied on driftwood brought to their barren shores from rivers draining faraway interior lands. This source of wood was widely dependable because the great rivers of Siberia—the Ob, Yenisei, Lena, and Kolyma —and the Yukon, Mackenzie, and Coppermine rivers of North America carry enormous quantities of trees washed from their banks each year during spring breakup. These are disgorged into the Arctic Ocean and carried throughout the polar basin. Where particular currents or storm waves prevail, jackstrawed piles of logs build so high and deep and wide that a man sledging on the sea ice can't come ashore with his dogs for miles. Conversely, other vast stretches of coastline are swept so bare by currents and waves that men can't even find enough driftwood to build a sled. In these regions, wooden sleds often were valued so highly that a man could purchase a wife on a one-for-one basis.

Wooden sleds were of two main types: flat-bottomed toboggans and sleds with runners. The toboggans worked best in soft snow that runners would sink into, but on packed snow or ice only a small running surface was needed, and runners were best. This meant that people who lived in forested regions, where snow stays fluffy, used toboggans. Those who lived on the windswept tundra, where snow generally is crusted, used sleds with runners.

Regardless of type, aboriginal hunters pulled their own sleds and when moving camp had their womenfolk help pull them. Dogs, which were small compared to modern Spitz, helped in hunting and maybe also acted as pack animals; quite clearly, they were eaten. But they weren't hitched as draft animals until late prehistoric time and weren't commonly used that way until the fur-trade era. It brought economic revolution, one effect of which was to make feeding dogs more feasible than ever before. Its culminating demand for pelts led to traplines as much as six hundred miles long, and men no longer could pull their own sleds. They needed too many supplies, and their traps produced too many furs to haul without help. Furthermore, using dogs to pull the sleds meant more speed, which meant more pelts and therefore more money. The money, in turn, meant more food for more dogs.

The toboggans of the boreal forest zone, as first used for man-hauling and later with dogs, typically were about two feet wide and ten to twelve feet long. Often they curved up on each end. Builders split boards from the green sapwood of birch or tamarack trees, then smoothed them with knives, and steamed and bent them. Next, using *babiche* (rawhide cut into long, thin strips), they stitched on crosspieces to hold the bed in place and then added flat running boards, also held with babiche. Each stitch was countersunk to avoid friction and wear in sliding against the snow.

On hard-packed snow or ice, toboggans are unmaneuverable and sleds with runners were the universal choice, varying in style from place to place according to purpose and materials. On St. Lawrence Island, the Eskimos of a thousand years ago man-hauled stubby sleds about fifteen inches long and fifteen inches wide to pull their skin boats and to carry meat and blubber from hunting site to village. Its runners were two walrus tusks. The Nivkh of Siberia had sleds as much as fifteen feet long and no more than a foot and a half wide. Teams of thirteen dogs pulled these. A driver wearing skis straddled the sled and steered with two sticks, which he also used for braking. Voice commands directed the dogs.

In spite of great environmental differences between regions, sleds of from eleven to fourteen feet generally were —and still are—favored throughout the north. Shorter sleds pound too much on rough snow. Longer ones are hard to turn in soft snow and unmanageable. Width typically is a foot and a half to two feet, about equal to the width of trail a man on snowshoes can pack most efficiently. The bed of most sleds rests six to ten inches above the runners, high enough to clear the surface of the snow and low enough to hold down the center of gravity. Upright stanchions line the sides of the bed, and a railing tops them. It projects beyond the rear of the sled to form handles for the driver to hold onto. Between stanchions a crisscrossing of rawhide or a latticework of short sticks keeps the load from falling out.

Construction can't be rigid. Sleds continually flex as they glide over irregularities, and their pieces must be lashed together rather than rigidly joined. This lets the sled "give" with the snow surface. A rigid sled would pound to pieces quickly. Runners are two or three inches wide curving up to the level of the bed, or higher, in front and extending out straight at the back to give the driver a place to stand.

Sapwood from a newly cut tree was (and is) best for the runners, just as for toboggans. It tends to freeze hard and provide a surface that can be glazed with ice, although this wasn't done in all parts of the Arctic. Steel runners work well at relatively mild temperatures, and some modern racing sleds even have a Teflon coating added to the steel. But when the winter temperature begins to hover below zero, most mushers replace steel running surfaces with hardwood, or switch to a sled with the appropriate surface already in place, since changing shoes is a slow job. In the old days, runners used to be shoed and iced by spreading a paste of mud or powdered moss, or even oatmeal porridge, onto the undersides, then spraying with mouthfuls of water, and rubbing the surface with the bare hand or with a piece of fur. Done right, this gave a smooth glaze of ice that lasted for a day or two and could be repaired easily by spraying again with water. Drivers carried a skin bag of water beneath their parkas constantly ready for emergency repair of the ice coating, or they used urine for the purpose. The shoeing bonded indefinitely to the wood. Only the glaze needed continual renewing.

A sled with wood runners that has been gliding smoothly across inland snow will stop abruptly when brought onto sea ice. This is because the overlying snow soaks up brine and, consequently, Eskimos who take their sleds onto the sea ice shoe their runners with strips of whalebone, which is unaffected by brine. The practice is both ancient and still in use today. This matter of precise snow conditions largely determines the ease—and even the feasibility—of sledding. Ideal temperatures are just warm enough for the pressure of the sled to melt the uppermost snow surface and provide a film of water for the runners to move on. In *Hunters of the Great North* Stefansson tells of asking an Eskimo companion why a certain journey would take ten or twelve days, when the same trip took far less time only a few weeks earlier. "Ovayuak [the Eskimo] answered that the weather was now much colder and that while a sled slides over the snow easily at such temperatures as we had in the fall, the runners now would grate on the sharp snow crystals almost as if we were dragging them over sand, and that we could not expect in midwinter to make much more than half the speed one could make in the fall or spring. He expected that both he and his wife would have to pull in harness with the dogs and, while he did not expect me to do this unless I felt like it, I must understand that he could not allow me to ride."

Wind and temperature both affect snow. Eskimos put off travel for a few days after a deep snowfall, waiting for wind to rework the surface into a crust and to drift loose snow over the rough places and smooth them. Such blessings can turn into liabilities, however. Often a breakable crust forms, or sastrugi turn tundra snow into a hopeless sea of sharp pinnacles. Nansen, sledging across sea ice to Franz Josef Land in 1895, wrote: "This is getting worse and worse. Yesterday we did nothing, hardly advanced more than a mile. Wretched snow, uneven ice, lanes, and villainous weather stopped us. There was certainly a crust on the snow, on which the sledges ran well when they were on it; but when they broke through —and they did it constantly—they stood immovable. This crust, too, was bad for the dogs, poor things! They sank through it into the deep snow between irregularities, and it was like swimming through slush for them."

Even when traveling on land under favorable conditions there can be moments when progress seems akin to making way through loose drifts of icing sugar. Once while on a dogsled trip along the Kobuk River, in arctic Alaska, we stopped to warm ourselves with tea carried in a vacuum bottle, and as I stepped from the sled I sank to my hips in the snow. We were crossing an open plain which in summer is a marshy quagmire and in winter a broad treeless expanse of white with just the tips of willows protruding. Just that fall men had set out tripods of saplings lashed together at the top to serve as route markers. The state highway department financed the project, which was the first "road" in the region. No construction was involved, just the markers, but routes previously had been recognized only by knowing the lay of the land. The snow ahead of us lay unbroken except for faint tracks left by previous snowmobile travelers, and our sled smoothed a clear swath into the snow behind us as a wake. When my feet strayed from this thin ribbon of packed snow, I dropped.

Dogs also flounder in soft snow, and a driver often has to go ahead on snowshoes to prepare a trail. He may even have to go over it again and again before it can hold dogs and sled. Freshly made tracks are soft, but by the following day the snow will have recrystallized and the trail become firm. To take advantage of this, drivers often showshoe a trail a day ahead of time. Even if it gets drifted over and doesn't show well, the dogs' paws can find it. So can human feet. You feel the firmness. But step off, and in you go. There then is no way simply to step back up. You must lean forward and start a half-swimming, half-walking motion, letting the inside of thighs and calves bear your weight at first. In this awkward fashion you can plow upward and forward fairly rapidly and soon regain purchase on the scarcely visible trail.

To keep the dogs pulling, our driver continually clapped his hands and whistled and clucked and chirruped to them. He kept them aware of progress by his voice, and held their interest in order to buoy their morale. Beyond question a dog team runs with high spirit. Our driver claimed that the hardest commands to teach are "Stop" and "Stay stopped." In the excitement of running, dogs reinforce each other's behavior and prompt still more running. Overcoming this is hard. One is likely to keep on going or to start up again after a short pause, and they all take off again. Unless dogs are tired, they want to run.

On warm days some mushers stop and let their teams eat snow to replenish body moisture. The dogs pant furiously during these rests, as they also do while running. This is their normal way of cooling off, although they also lose excess body heat from the thinly furred skin of the groin and "underarms." On extremely bitter days the Yukaghir and Yakut of Siberia wrap their dogs' groins with soft strips of skin as protection from the cold. To keep their dogs' paws from getting cut on the sharp edges of snow crystals, women in some parts of the Arctic make little shoes of sealskin or reindeer hide for the dogs, cutting a hole for the nails of the two front toes to stick out through. Careful drivers watch the snow and put these booties on as soon as it becomes abrasive. Others wait until they notice bloody footprints. Traveling through a wet blizzard it sometimes is necessary to stop and brush snow from the dogs' eyes. Otherwise their faces get so covered with slush that they can't see.

Even when all is going well the chances are it won't last long. There may be no snow falling to blind the dogs, or rough snow to cut their feet, or crusted snow to break beneath their weight, but almost surely, sooner or later, two dogs will start to fight, or the whole team will chase off after caribou or a moose. If not this, the snow may well need to be packed ahead by snowshoeing, or the sled will burrow its nose into a snowdrift and have to be jerked free. Once everything is ready to go again, the runners may stick and need vigorous rocking and lifting to overcome their bond with the snow and get the sled back in motion. Not even passengers simply tuck into the folds of caribou robes and ride for hour after hour. Nothing muscle-powered can be that simple. Besides, days cold enough for good sledding are too cold to sit for long. Dogs can haul sizable loads in reasonable time, however, with distances, speeds, and weights varying a great deal. Probably a load of around seventy-five to one hundred pounds per dog is representative and, given that weight and pulling under reasonably favorable snow conditions, a nine-dog team can travel thirty to forty miles a day for weeks at a time.

Dogs are harnessed in different ways. The eastern Siberian method seems to be the oldest, a simple bight of the trace placed over each dog's head. This means pulling with the neck, and excitement or extreme exertion tends to choke the dog. To avoid it, people along central Siberia's Amur River placed the bight around one front leg as well as the head so that the pull came from the shoulder and part of the chest. On Sakhalin Island, off the east coast of Siberia, dogs pulled from the hips, and their tails were cropped to avoid interference with the harness. Two main methods of hitching to the sled itself traditionally were common: Either the dogs were strung out along one line, or they were hitched fanlike by several separate lines which might be as much as fifty to sixty feet long (but usually were shorter). In forested areas the one-trace method worked best, with the dogs attached along it single file on alternate sides, held by collars that usually were fitted with swivels of ivory, antler, or bone to prevent twisting. This staggering kept the sled going straight. On open tundra or sea ice, the fan hitch gave the same evenness of pull. Hitching in pairs with a lead dog in front seems to be a recent style. It lets the feet of the dogs pack the trail for the sled runners, minimizing need for the driver to snowshoe ahead.

The process of getting a nine-dog team unchained from the stakes and hitched to a sled takes a person working alone from fifteen minutes to half an hour if all goes well. On return from a trip the process has to be reversed, and the dogs fed. One of the main advantages of snowmobiles over dogs is eliminating this time spent hitching and unhitching. With a snowmobile the only preliminary to getting underway—usually—is to pull the starter cord. On return all that is needed is to stop and climb off. There is no laying out of harness or putting it away, no chaining and unchaining of dogs, no fights, no time and labor spent in feeding.

Another reason many arctic people have switched to snowmobiles is that pouring gas into a machine is easier than getting food for a dog team, then storing, preparing, and dispensing it. A sled dog weighs about fifty pounds and needs half as much protein per day as a man does. Multiply this by keeping five to ten dogs, and it is obvious that as much work goes into fishing and hunting for a dog team as a family. Two or three tons of flesh may be needed each year, and even in summer when dogs aren't being used they must be fed.

Most of all, ease of travel and speed have led arctic peoples to abandon dogs for snowmobiles. A man on a snowmobile can go two or three times as far in the same number of hours as he formerly could by dog team. He also can pull a longer, heavier sled and do so under a greater range of snow conditions. He seldom even has to snowshoe ahead to pack a trail. In soft snow all he need do is uncouple the sled and roar out on his machine, then come back after a double pass over the trail and hitch on the sled. Furthermore, while he is doing this there is no chance of his dogs running off.

The snowmobile's power and speed mean three loads of firewood hauled in the time it takes for one with a moderate-size team of four or five dogs, and it means hunting caribou thirty or forty miles away and returning home the same day instead of staying out for two or three days. No tent need be taken along, no stove and saw and sleeping robes, no dog food. The likelihood of drastic weather change while on a trip is minimized. The full capacity of the sled is available for bringing back caribou. With snowmobiles, even men who work for wages through the day can subsist partly off the land by hunting and fishing after hours and on weekends. Trappers lose less fur to owls and ravens and foxes because they can check their lines more often, and they don't need to spend time in making food caches for dogs or for themselves. Since the snowmobile easily keeps pace with fleeing caribou, or moose or polar bear, hunters can shoot at closer range than before—and the result has been a decline in some wildlife populations, especially caribou and to some extent polar bear, although in their case trophy hunting by airplane probably was even more directly responsible. In places lakes that are within day-run distance of villages have been fished out, but since this almost surely would have happened anyway now that fixed settlements have replaced the old seasonal nomadism, the machines aren't wholly to blame. Furthermore, snowmobiles unquestionably have saved millions of fish and thousands of caribou once fed to dogs.

The machines haven't totally simplified life, however. They break down and wear out, often on the trail. Usually first to go is the ignition system. Soon not even changing the spark plugs and priming with starting fluid will get the motor going. Dogs took time to get underway, but always started. Also, if one dog gave out, the others still pulled. But let even a small part of a snowmobile break, and the whole machine may be useless. Temperature alone can have disastrous effects. At −30° F. and lower, rubber drive tracks turn brittle and may break; yet trips for stove oil, or wood, or drinking water are needed, and men drive their snowmobiles regardless of the cold. Damage is common. At these temperatures starting can be trouble—whether starting up to return from a church meeting or a distant hunt. It may be necessary to set spark plugs around the burner of a gasoline stove to warm, or to hold a blowtorch close to the block, or set a portable stove close to a snowmobile covered with a tarp.

Frostbite injury is far more common among snowmobilers than dog mushers because the machines' speed creates a chilling wind that drivers steadily face into, whereas handling dogs means getting on and off the sled, riding awhile and running awhile. Even fingers move constantly while sledding but not when gripped around snowmobile handlebars. The noise of snowmobiles is another disadvantage, so significant that hearing loss has become common in the Arctic. The concentrated weight of snowmobiles also is a problem. They break through weak ice and immediately sink, whereas dogs generally avoid such places instinctively, and they keep pulling if caught.

Nonetheless, snowmobiles' advantages so markedly outweigh drawbacks that acceptance has burgeoned throughout the north. The Kobuk Eskimo village of perhaps fifty households that we visited got its first snowmobile in 1965. Eight years later only one family relied exclusively on a dog team; everybody else had a snowmobile. Throughout the Arctic, the machine revolution came with equivalent suddenness. For instance in northeastern Finland the first Ski-Doo arrived from Canada in 1962 and was purchased by a bachelor schoolteacher, who planned to use it to fish recreationally but soon was hauling wood and groceries with it. The next winter a forest ranger and a nurse bought snowmobiles; so did a postmaster and four relatively wealthy Lapp reindeer herders. By 1967—only five winters after arrival of the first snowmobile—all the herders of the district had one. Many reindeer herders seem to have felt the snowmobile revolution even more keenly than traditional hunters, fishers, or trappers. As illustration, herding practices among the Skolt Lapps of northeastern Finland have changed so drastically in response to mechanization that a high percentage of families no longer own deer, which previously provided transportation, food, and cash potential. Pertti J. Pelto, an anthropologist specializing in sociocultural change, terms what has happened a "reindeer disaster," based on the years 1957 to 1971 when snowmobile use soared from rumor to commonplace.

The Skolts are a small group of Lapps who lived under Russian domination from the mid-fifteenth century until World War II, when they fought on the side of the Finns and decided afterward to resettle in Finland near Lake Inari. In their former situation individual families closely controlled their deer. Herds were small and not allowed to migrate in the manner of Lapp reindeer elsewhere. Constantly moving with their deer wasn't practical for the Skolts because they relied on fishing as well as on herding. They couldn't afford to let herd impulses govern where they were at any particular time; the seasonal demands of fishing had to take precedence. Consequently they tamed their reindeer herds. Even in summer when the deer were grazing on their own, they stayed close to men, drawn by smoky fires specially built to ward off the worst of the mosquitos and flies. Any genetic urge to wander that still lingered was controlled easily by slaughter. Ancestral migration patterns thus disappeared, and the herds' ranges became restricted by the deers' own man-induced inclinations.

The large quantities of meat and hides regularly available from the herds brought cash for newly obtainable material goods, including food staples such as flour, sugar, coffee, and tea. Hitched to sleds, geldings were useful not only for hauling firewood and drinking water, but also for traveling to the trading posts. Three types of sleds prevailed: a high-stanchioned one pulled by three reindeer hitched abreast one another, used for long-distance trips across relatively open country; a smaller runnerless sled shaped almost like a boat and pulled by a single deer, frequently used in the hilly back country; and a stubby low-stanchioned sled for heavy freighting. It too was drawn by a single reindeer, often with an extra one brought along as spare. Skolt songs boasted of draft geldings with slim necks and beautiful antlers running easily through the snow even "when the drifts over their knees reach."

When they first moved to Finland, Skolts fenced their

reindeer to keep them from trying to go back to home territory. But in five years the herds grew too large for continued winter grazing within the fenced area, and owners turned their animals loose. From then on the reindeer roamed free all summer. In early fall, owners individually captured their tamest geldings for pack and sled use. A few weeks later, when the snow cover was right for easy travel, men began working together to round up all the deer and drive them to a central corral for sorting into the herds owned separately by families. Men on skis drove the animals, aided by dogs skilled in coaxing along stragglers and chasing hideouts from their brushy thickets.

Ear notches let owners identify their own herds and, once the animals were corralled, each man lassoed his deer, slaughtering some for family use, selling others to meat buyers, and gathering the rest to tend through the coming winter close to home. The herding seemed almost like a harvest as each man tallied the return of his animals, noted the reappearance of any that had evaded the last roundup, and counted and claimed the new calves accompanying his cows. The gathering and sorting lasted for two or three weeks, giving everybody time for both work and celebration.

Additional roundups continued through the fall, until by early February tundra and forest had been cleared and almost all deer were again in the direct charge of individual owners. Through the winter families protected their herds from predators, butchered a few deer for meat or to sell for cash, castrated bulls, and worked at taming geldings for draft use. As spring calving drew near they easily caught and tethered pregnant cows, moving them from place to place for grazing. That way newborn calves could be tended and earmarked in the first few days after birth, then turned loose with their mothers. By early June all the reindeer were again roaming free. The relationship between men and deer was close, the pattern clearly established, the cycle self-repeating from year to year.

Then came snowmobiles, and everything changed. In just one decade the total number of reindeer owned by the Skolts dropped by a third, from twenty-six hundred in 1961 to fewer than seventeen hundred in 1971. Draft animals disappeared. Most of the families who had relied on deer no longer owned enough to be of real significance. The high cost of herding had driven them out. Those who still had herds were in touch with them only during roundup—and it wasn't the same. Men no longer spent weeks on skis gathering the deer and living with them as they drove them to the corrals. Instead, speeding along on snowmobiles, they rounded up only small bands at a time and lost a high percentage as the drive progressed. The noise frightened many deer into bolting for cover, and once within the brush they were hard to rout out, since snow around trees or willow thickets is too loose to operate snowmobiles in. Bands dwindled still more as the drive progressed because small groups of reindeer have little "herd pull." They are easily drawn off by wild herds.

Consequently, roundups grew smaller, and the working relation between men and deer almost disappeared. Panicky, pell-mell fleeing from snowmobiles cut the weight of deer ultimately corralled 20 percent from what it had been in the days of gentler roundups. Cows began to drop fewer calves. The stress of mechanized herding seemingly left some deer crucially weakened, unable to muster energy enough to feed well through the winter and carry pregnancies to term. Furthermore, winter herding no longer was practiced and therefore a given cow might be repeatedly rounded up and repeatedly stressed. High calf mortality resulted. Skolt families that had been self-sufficient were forced to choose between dole and emigration to the outside world. Formerly "starting capital" in the form of deer was given to children on occasions such as christening day, birthday, and so on, and by puberty each young person had his or her own nucleus herd. But no more. With the herds gone, capital now can come only from the outside. The pattern of husbandry has changed totally.

The practice of reindeer herding stretches from Lapland, in northern Scandinavia, across all of Eurasia to the Chukchi Peninsula and Sakhalin Island, but seems never to have crossed to America (until the none-too-successful introduction among Alaskan Eskimos in the 1890s). Seemingly this is man's most recent domestication, occurring around two thousand years ago. Annals of the Chinese Liang dynasty, A.D. 502 to 557, tell of the land of Fu-sang (probably the Lake Baikal region of Siberia) where deer were raised for their milk; and T'ang dynasty records, A.D. 618 to 907, speak of Ku, southeast of Lake Baikal, where "in like manner as cattle and horses are employed in China, the reindeer are used here for drawing sledges, which are capable of carrying three or four persons. The people clothe themselves with reindeer skins. The reindeer subsist on the moss of the soil."

Almost universally, reindeer people catch their animals by lassoing them, and they castrate bulls by biting the scrotum, mashing the spermatic cords. Other herding details differ. In former days the Kirghiz of the Siberian steppe and the Tungus along the Yenisei River used dogs to guard their deer from wolves at night, although Tungus peoples elsewhere didn't do this, and neither did other tribes farther to the east. Most Tungus trained their dogs to help in hunting by overtaking wild deer and circling to detain them until marksmen could arrive. Lapps also used dogs to circle after deer, but in their case it was to drive domesticated stragglers back into the herd.

How did it start? Some say domesticating reindeer was an outgrowth of keeping cattle and horses in the Lake Baikal area, as tribes simply added deer to their other herds. Others say the beginning probably was on the headwaters of the Amur River. Tungus there would have observed domesticated camels and yaks, as well as cattle and horses, and when they subsequently pushed northward to the Yenisei region they may have taken along cattle and horses and substituted reindeer for camels and yaks, which couldn't adapt to the new environment. Another possibility is that Samoyeds were the first to adapt herding techniques to reindeer, bringing the practice north and introducing it eastward to the Tungus and on to the Yakuts, Chukchis, and Koryaks, and westward to the Ugrian people of the Urals and their linguistic relatives, the Lapps.

Or it may be that none of this led to domestication of

reindeer. Tending deer may have grown out of hunting them. Certainly an element of wildness remains in the tamest reindeer, and much of man's relation to them is more like hunting than usual herding. In pre-snowmobile days Lapps spent days just in searching for scattered herds, following footprints in the snow or tracing the deer by the direction trampled vegetation pointed. In corralling they drove the herd into a funnel-shaped enclosure, much as early hunters the world over drove their prey, from rabbits to buffalo to fish. Initially, a salt lick of human urine in the snow lured wild reindeer, and it continued as reward for domesticated animals. Lassoing probably came directly from hunting. Early-day Lapps and Samoyed hunters captured wild deer with lassos; Chukchis lassoed mountain sheep.

Probably man's first advantage in gaining mastery over deer was to use those he tamed as decoys. Until about the time of World War I, when their cultures were disrupted, Tungus and Samoyeds trained decoys to graze at the end of a long line close to wild reindeer. Holding the other end of the line, a man in hiding could slowly draw in his tame deer and thereby lure the wild ones within easy shooting range. Another technique was to tether tame females for wild males to mount, or to tie thongs around the antlers of tame bulls and allow them to mingle with wild cows. Inevitably, this would lead to fighting with wild males, which would get their antlers tangled in the cord and could be shot. At the outset, according to this concept of domestication, the main value of spending time in taming deer was to insure success in hunting wild deer. Once that relationship had begun, other aspects of herding were borrowed from cattle and horse cultures. Reindeer used as saddle animals and for packing seem to have derived mostly from Mongolian and Turk horsemen, although Lapps may have acquired saddles from Scandinavians.

Pulling sleds with reindeer probably was adapted from dog-sledding. Reindeer harnesses show no similarity to those used with horses or oxen but are like Siberian dog harnesses. Deer have advantages over dogs in greater strength and also in their self-sufficient manner of feeding. "As soon as the wind blows a little, the dog cannot travel," reads an early report from Alaska. "Especially this is so if the wind happens to be in the face. [On the other hand] the deer does not mind the wind in the least, from whatever direction it comes, it rather enjoys traveling against the wind. It costs nothing to feed, it faces all weather . . . and trail or no trail it will haul its two hundred pounds or more day after day, week after week."

Reindeer herding remains the economic key in much of Eurasia. In 1973, newspapers reported the plights of Chukchi herdsmen whose 370,000 reindeer were kept from feeding by freakish ice conditions. Planes, helicopters, and dogsleds were being used to haul feed and salt to the starving animals and to search out ice-free pastures where they could be driven. Reindeer and their herders across a northern swath of Scandinavia constitute Lapland; it has no other territory, no political entity. Lapland is a tradition more than a land: By consent, those of its people who wish to do so still follow their herds from the Norwegian coast and islands, where the deer summer, to the woods of Sweden and Finland, where they winter.

In Alaska reindeer herding never really has caught on, although it was introduced there nearly a century ago through the persistence of Sheldon Jackson, a Presbyterian missionary who in 1885 was appointed an education agent. He fervently sought to improve the Eskimos' lot, viewing the simplicity of their life as cultural poverty. "A change from the condition of hunters to that of herders is a long step upward in the scale of civilization," Jackson wrote. "While we offer the Gospel with one hand, we must offer them food [reindeer] with the other." Eventually the government authorized importing a nucleus herd from Siberia, and in 1892 the American flag at last was raised over Teller Reindeer Station, named for a U.S. senator who had supported the project. First Chukchi herders were brought to teach the Eskimos care of the deer, then Norwegians and Lapps.

For three summers the revenue cutter *Bear* was so involved in locating and transporting Siberian reindeer that its other work had to be neglected. In his annual report for the years 1895–96, Jackson wrote earnestly: "It is now found that the reindeer are as essential to the white man as to the Eskimo." Reindeer were hauling goods to "the wonderful placer mines of the Yukon River," he reported, and the Army had sent "tents, rations, and camp equipage" to Nome by reindeer sled when troops were called in to quiet disorder during the mining boom there. Deer had replaced dogs on several Alaska mail runs and, along the Yukon where men were building a telegraph line, they had replaced mules which were unable to cope with snow.

Jackson's dream seemed vindicated. By the late 1890s a total of 1,280 reindeer had been imported and distributed. Breeding had succeeded. Sled use was well begun. Additional reindeer stations had been established. But one vital aspect still was missing: The Eskimos showed little interest. The reasons were many. Reindeer unquestionably could thrive and be useful in Alaska, but the demonstration had problems. Herds were too small to generate the centrifugal force needed to hold them together; they joined wild herds and were lost. Parasites and wolves took a toll. Warble flies riddled hides and ruined their value. Overgrazing destroyed lichens for miles around the stations. Freight charges for shipping meat to distant markets were high, and storage of any quantity of meat following slaughter was difficult. Also, it struck Eskimo owners that tending reindeer gave little satisfaction. Its only hope was for a return at year's end, whereas hunting yielded immediate results.

In 1939 the U.S. government bought up all the deer. Defense spending was rising, and the demand for military parkas put a premium on the skins of calves. Herding was reorganized to bring predator control, range management, and constant close herding. This succeeded enough for the reindeer industry to continue in Alaska, but even today it hasn't grown much or been widely adopted. Perhaps growth and greater acceptance will come under management by native corporations, which have just entered the industry. So far, most Eskimos continue to prefer wild caribou to tame reindeer, a tie to hunting that still contributes to their live-lihood and recently has received legal recognition for both Eskimos and Indians in Canada and Alaska as part of native-claims settlements.

Hunting methods vary throughout the Arctic but always have centered on the animals' migration. Stone cairns set in a row fifty yards apart for five or six miles formerly would lead to a lake where women and children waited along the shore and men on the water in kayaks. When the caribou approached drifting along the line of cairns, the women and children howled like wolves, or leaped from hiding, to stam-pede them into the water or onto thin lake ice. There the men speared them. Or, instead of leading to a lake, some fences funneled into a rocky defile or a river crossing where hunters waited. In forested country, poles set upright served instead of piled stones. Deadfalls also were used widely, usually with drift fences leading to them. Eskimos built pits entirely of snow. They mounded it up with an easy incline for walking to the top of the mound, and an abrupt drop into a pit that was concealed with snow and moss. Often they set knives upright into the snow at the bottom. An 1895 report on the inland Caribou Eskimos of northeast Canada even mentions weighted snow slabs set on wooden axles to drop caribou one at a time into a pit. The slabs could bear the weight of an animal until it got past the middle; then they would trip, and spring back ready for the next caribou. A sprinkling of urine on the snow acted as bait.

In Siberia earthern pits as much as 130 feet on a side were common for catching wild reindeer. Central Ural families in the late 1800s owned from 50 to 300 of these pits apiece. The Transbaikal region had more than 20,000 deer pits in thirty-six hundred square miles, with nearly 2,000 miles of drift fences leading to them. The toll of animals taken in the pits mounted so high that the deer learned to avoid them by ceasing to migrate. The pits thus actually exerted a selective genetic influence, according to Russian biologist A. A. Nasimovich.

Similarly in Alaska, reports Bob Uhl of Kotzebue who has hunted with Eskimos all his adult life, snowmobiles have prompted at least behavioral change, whether or not genetic. Just the sound of a motor sets caribou fleeing and, in the few years since snowmobiles' arrival, they already have learned that the machines can't follow up a steep slope, into a rocky place, or across snow-free ground. Only those that grasp this truth survive. Herd animals can learn from experience in a way that solitary animals such as moose or bear cannot. For these creatures the first chance to learn what happens when a snowmobile gets close may well be the last chance. But with caribou some survive even a considerable slaughter, and a leader that has learned from the experience will be followed as later snowmobiles approach. Caribou now even can discriminate between the mail plane's engine and a snowmobile's. They become alert when the plane passes over, but if the whine of snowmobiles comes into range, they are off.

The efficiency of the snowmobile, and earlier the rifle, has brought waste. With a convenient and assured power to kill, hunters have overkilled and seemingly mostly from this one cause caribou herds have declined. Canadian barren-ground caribou once numbered two or three million but by 1955 biologists counted only 278,900. Today the count is back up to half a million. In Alaska the arctic herd, which stayed at about 250,000 caribou, has dropped to only about 60,000. For the barren-ground herd the rifle is largely to blame; for the arctic herd, the snowmobile. Other factors may be involved, as well. Perhaps man's battle against wolves contributed to an unnatural excess of caribou, which now has triggered a built-in population adjustment. Probably sport hunting has taken some toll, and quite possibly the stepped-up human activity brought by arctic-slope oil and gas development may have had an effect already, even before the pipelines have physically influenced migration routes. Maybe the dieoff is in part owing to some little-recognized natural cycle. Hopefully, restricted hunting will let the caribou rebuild whatever the reason for their decline. The traditional arctic triumvirate of men, dogs, and deer never again will be as it was; times have changed. But, if all goes well, it will endure and even prosper along new lines not yet known.

CHAPTER

10

SKIING

Not only in the Far North but throughout the snow belt of North America and Europe, the two-cycle whine of the snowmobile became the sound of winter through the 1960s— a sound in the late 1970s or the 1980s perhaps destined to be at least partially silenced by the energy crisis. Where man once depended on sledding, snowshoeing, or skiing, his technology now provides effortless zipping about by machine. No training is needed to operate a snowmobile, no particular coordination or license, not even a road. Just go. Snowmobiling is easy, fast, and fun. It answers an inborn urge.

Northerners have come to rely on the machines for winter transportation, law enforcement officers and wildlife wardens make their rounds by snowmobile, and children are "bussed" to school by them. Millions of dollars go into snowmobiling as recreation. Alaskans even air-freight their machines to enter the 160-mile race from Nome to Teller and back. Contestants in it have careened across the land as fast as 90 miles an hour despite temperatures of − 45° F., thus creating their own awesome wind-chill factor. The effects of literally millions of snowmobiles unleashed upon the winter world are legion—and notorious. Wildlife suffers from disturbance and from mechanical alteration of snow cover, as well as from increased vulnerability to hunting. Plants get broken by passing machines, crushed by their weight overhead, and damaged by cold, since compacted snow offers little insulation. Snowmobiles break through the ice of frozen lakes and rivers, bringing death. They are costly and noisy, wholeheartedly disliked and equally wholeheartedly welcomed. For many who live where winter is long and snowy the machines have brought welcome revolution; for others they have brought an unconscionable desecration of winter purity. For better and for worse snowmobiles have done to today's snow country what the tin lizzie automobiles of yesterday—also for better and for worse—did to greener lands and in milder climates. The human impulse is to go and to do, and snowmobiles make it easier in winter.

Yet the machines aren't the first device to facilitate travel over snow. Man has had proper equipment for millennia and with it long ago divorced himself from dependence on the sea for travel and for food even in the snowbound north. Wearing snowshoes or skis he learned to walk along river valleys and across bogs when winter brought its yearly gifts of cold and snow. Frozen rivers then become highways to travel with equal ease in either direction, and the bottomless ooze of boggy ground no longer acts as an obstacle. Even forest travel is simplified once brush disappears beneath the snow. The long story of transportation has no more important chapter than that concerning the invention of snowshoes and skis. They underlie northern culture with one major exception: Most Eskimo peoples stayed along the windblown arctic coast where snow is crusted and firm, and their winter hunting was for seals and walruses. They traveled across the sea ice, not the land, and didn't need elaborate gear for their feet. Some groups had crude snowshoes but on the whole Eskimos had no such basic reliance. Most other northern peoples did.

Curiously, southern hemisphere peoples had no snowshoes or skis. The Ona Indians of Tierra del Fuego sometimes tied small bundles of bushy twigs to their moccasins when walking in fluffy, new-fallen snow and their name for this translates as "shoe snow." Their whole approach to the problem was crude and temporary. Andean mountain people similarly developed no notable snow technology. That distinction belongs to the northern hemisphere, where it is fundamental from Japan across Asia and Europe to America. In general wooden-plank footgear, or skis, belongs most fully to Eurasia and webbed footgear, or snowshoes, to North America. Probably the earliest use of each was in central Asia, although evidence is skimpy. Wood and fiber webbing seldom last long and even archaeological traces can be expected only from dry caves or continually wet mud where chance preservation is possible. Regardless of origin, however, the practice of enlarging the feet to ease travel over snow is a natural concept for inventive peoples and individuals, and the fact that a few snow-country groups lack this technology is perhaps more remarkable than that most have it.

Snowshoes probably came first. In simplest form they are easier than skis to make and to use, whether no more than a tree bough added to the feet or a crude frame with some sort of crisscrossing. Snowshoes were used by early farmers in Sweden and Norway and pastoralists in northern Spain. They are known in ancient Tibet, on the Amur River, and among the Ainu of Sakhalin and Hokkaido. In America they range from the Atlantic to west of the Rockies, and from Arctic to desert. Frames were of everything from willow to spruce to whalebone. Shapes were rounded or oval or hourglass, with a blunt toe and pointed heel, or pointed at both ends. Some European styles even are ladder-shaped—two straight boards connected with crossbars and left open at toe and heel, a stiff and clumsy contrivance but better suited to snow than the mere booted human foot. Webbing also has varied widely. Best is babiche, but sinew or even fishskin will do. Ainus and Mackenzie delta Eskimos simply used six crossboards and practically no webbing, and Mesa Verde Pueblo Indians sandwiched yucca leaves and roots between crude twig frameworks.

Even within a tribe, different snow situations called for different styles. Northern Athapascan snowshoes worn for hunting and trapping are four and a half to five and a half feet long and eight to ten inches wide. Trail shoes are narrower and only about three feet long. They are more maneuverable and because they sink deeper into the snow are better for breaking trail ahead of a dog team.

The oldest known wooden skis belong to late Neolithic times, four thousand to five thousand years ago. They are of two types, arctic and southern. In the earliest southern style, thongs pass through holes in ridges along the sides of the footrest. Tied to the foot, these thongs hold the ski on. This design was poor, however, because even in moderate use the side ridges broke easily. To correct this, men soon started thickening the footrest and passing the thongs through it horizontally, then bringing them up around the foot. Skis of this type have been found in bogs from the Ural Mountains to Norway.

Arctic skis had no raised footrest. They were held to the feet by thongs strung vertically through the skis themselves, and sometimes the bottoms of the skis were grooved between the holes to accommodate the thongs, so that they wouldn't interfere with sliding and would wear well. The oldest known skis of this type come from Sweden and have been dated at 2000 B.C., and, since time would be needed for such refinements to develop, these particular skis must have had forerunners unknown today. The Holmenkollen Museum on the outskirts of Oslo displays several ancient skis discovered in Scandinavian bogs, six of them older than one thousand years. In style and proportion they look about like the wooden downhill skis of the 1950s.

The earliest written record of skiing is a petroglyph chipped into the rock near Rödöy on the west coast of Norway. It shows a man skiing, his feet shod with long boards upturned at the toes. The silhouette is unmistakable and could as well belong to a modern skier as an ancient one. The skis are perhaps a little long even for cross-country skis today and the toes more sweeping, but not much. Skiing is one of mankind's oldest modes of transportation, and basically it has changed little. With the early skis at Holmenkollen are ten ski quotations from the years A.D. 920 to 1120. Even older written mentions are known in the seventh-century annals of China's T'ang dynasty. These speak of fishing and hunting peoples who wore "wooden horses" on their feet and used "props" under their arms. Thus wondrously equipped the people "went forward at least one hundred paces with every stride."

As is true of snowshoes, the style of skis varied from place to place according to differences in snow and in human need. The widest and shortest were developed by the Siberian Evenki people who live in a forested zone of notoriously fluffy snow. The Holmenkollen collection has a pair of these only four and one-half feet long by ten inches wide with the toes only slightly bent up. Among the Nivkh people at the mouth of the Amur River and on Sakhalin Island the mark of a good hunter was to foretell accurately which type of skis to use according to season and weather. Fairly long, plain board skis were fine for winter trips to cut wood, for fishing through the ice, or for visits to villages not too far distant. They worked well in deep snow or on spring crust. But for hunting, where speed and silence were important, skis lined on the bottom with fur worked much better. Men attached strips of hide layer on layer along the whole length of the ski using fish glue, and sometimes they added whalebone edging. Skin of moose, deer, and seal worked best, each for a specific use in a certain type of snow. Linings of moosehide slid poorly on winter snow but were fine in the warm fall or spring weather, and were particularly valued for sliding on the granular snow of spring without making noise. Deerhide slid smoothly on ice or snow in cold weather; skis lined with sealskin were equally usable but wore out sooner. A man needed to have a pair of each kind to live successfully. Ski poles were flattened into small shovels at the top, useful for cleaning snow off the skis, digging traps free, burying automatic-release projectiles used in hunting, and for cutting snow blocks to hide such devices.

In North America skis remained crude when compared to those of Eurasia or to the continent's own highly developed snowshoes. The concept was distributed fairly widely in aboriginal times but never refined. Modern skiing came to North America as a European import little more than a century ago. The first colonists came mostly from England and

France and Germany where skiing was little known, but when Scandinavians began arriving in America they brought their winter "long shoes" with them.

Today winter and skiing go together so naturally that even those who know the sport only through television have difficulty realizing how novel it was to most people just recently. The pioneering snow surveyor James Church wrote that "because of the speed and sport value, skis have become the dominant showshoe although their use is exacting a high toll in accidents," mostly because of too tight "lashings." Webbed snowshoes had been developed in Canada, he explained, "where powder snow and thickets abound," whereas Norwegians had been the ones to perfect skiing "on wooden skates because of [their] firmer snow and open country which afforded opportunity for maneuvering and for high speed on down slopes." American skiing still was rare enough in the early 1900s that such explanation was enlightening. Best estimates are that even by the mid-1940s there were only about two hundred thousand skiers in the United States. The figure now is ten million, and rising.

Scattered beginnings must have introduced modern skiing to the Americas but records are scarce. A Colorado man named Father Hyder apparently was skiing dual rounds as preacher and mail carrier as early as 1861, and ten years later a Montana miner named A. Bart Henderson skied routinely between work and town. Henderson was laying out a toll road through Yellowstone Canyon (now Yankee Jim Canyon), and he wrote of continuing work until Christmas Day, then going into Bozeman "on a pair of 15-foot snow shoes." A note in the November 28, 1883, Livingston, Montana, *Enterprise* tells of a man named Frank Woodcock who used skis to carry the mail "and met with hardships . . . [that] would have killed."

East of the Rockies skiers seem to have taken to New England slopes for sport by 1880, and a ski club was organized in Minnesota a few years later. In Canada modern skiing probably first began in 1881 at Montreal. Not much later skis were introduced to South America by the Chilean and Argentine governments. They had employed "some three hundred expert skimen imported from Sweden and

Norway to take the mails from Valparaiso to Buenos Aires."
A 1901 article in *Outing* magazine describes these stalwart
mail carriers as "snow skaters" and their skis as "long tough
wooden shavings . . . fifteen and even twenty feet in length
[with] the forward point tapered slightly and curled up like
the nose on an old-fashioned Dutch skate."

Before these new-world beginnings California had experienced a ski craze—and it wasn't the legendary Norseman
known as Showshoe Thompson who first started it, as commonly is believed. Thompson made his first trip skiing mail
from Placerville, California, to Carson City, Nevada, in January of 1856, a ninety-mile four-day feat with a heavy ruck-
sack of letters and packages on his back. But six years before
that skiing had begun in the Sierra gold fields at Rabbit
Creek, a boomtown now known as LaPorte.

Old-timers of a generation ago talked of depths of thirty
feet on the ground there by February or March, and this
sounds accurate in light of contemporary measurement. To
live in such snow they entered their houses through sloping
snow tunnels dug to attic windows, and they spliced their
chimneys in late autumn to lengthen them and be sure they
would reach above winter's snow. Leaving home in the morning they lashed poles with bright colored flags to their roof-
tops as beacons to head toward after work. But not even that
always worked. Winds often buried chimneys beneath snowdrifts or blew down the poles and finding home then became
a real problem. About the only thing the deep snow didn't
affect was employment, which was underground in the
mines—or at saloons and brothels, accessible on skis no matter how deep the snow. The first year's skis, fashioned of
barrel staves, evidently were the idea of Scandinavian sailors
who had jumped ship in San Francisco and joined the stam-
pede to the mines. Soon everybody had a pair, with the
initial design quickly improved upon. "Long snow shoes"
were standard winter equipment throughout the district
from infanthood into dotage. Charles Hendel, a mining
engineer and surveyor who arrived at the mines in 1853, still
was skiing thirty-five miles into Quincy for county super-
visors' meetings at age ninety-six.

Perhaps as early as 1853, and certainly by 1857, LaPorte

was hosting downhill ski races. Each town in the mining district had its own club, and there never was a problem in attracting contestants to line up for prize money collected mostly from local saloons. Each club would announce the date of its annual race in early winter, an exciting period as notices went up on the bulletin boards of post offices, general stores, saloons, livery stables, and blacksmith shops. Rules were simple. Skiers stood in a line at the top of a slope, someone hit a circular saw with a hammer, and at that signal a flagman dropped a red kerchief to signal timekeepers at the finish line to start their stopwatches. When racing hit its peak in the late 1860s and into the mid-1870s, contestants used skis ten to fourteen feet long, about four and one-half inches wide, an inch and a half thick under the foot, three-quarters inch at the back, and one-quarter inch at the front. With these strapped to their feet they responded to the gong by shoving with their poles and striding to get up speed. They bent to a squatting position to cut wind resistance, held their poles parallel to the snow, and schussed. If both the snow and the choice of "dope" applied to ski bottoms were right the men shot downslope at speeds reported as eighty to ninety miles an hour, a speed that sounds high but may be accurate considering the nature of the race courses. Even Snowshoe Thompson is said to have been taken aback when he accepted LaPorte's invitation to compete in the race of 1869. That year it was on Lexington Hill, and Thompson found himself at the top of a steep two-thousand-foot chute with all trees removed. The snow was slick, and he hit an icy spot and fell soon after shoving off. On a second try he kept his balance but finished the race last. Humiliated, Thompson invited his hosts to come to Silver City and race under the rules of his own Alpine Club. This meant running against time and over jumps for four miles through heavy timber. LaPorte skiers never took up the challenge, perhaps out of recognition that downhill and cross-country skiing are altogether different.

That same year of 1869 marked Snowshoe Thompson's last run through the snowy Sierra on his official rounds. The railroad and stagecoaches were taking over. Thompson had been born April 30, 1827, as Jon Torsteinson Rui. His family lived in the mountainous Telemark province of Norway but moved to the United States in 1837 to farm. They settled first in Illinois, then Missouri, and finally Iowa. The California gold rush drew young Jon west, a six-footer in his early twenties. He tried his luck but found neither wealth nor satisfaction. He changed his name to Thompson, easier to pronounce, and switched from mining to farming. In 1855 he happened onto his unique calling. An autumn issue of the *Sacramento Union* carried a headline proclaiming: "People Lost to World! Uncle Sam Needs Mail Carrier." The new farmer sensed opportunity. He loved the mountains and knew that he had a good sense of direction. He remembered childhood days when nobody was stopped from their travels by snow because everybody skied. Why shouldn't he try skiing across the Sierra to tie communities isolated by snow back to the world that had "lost" them?

First step was to make a pair of ten-foot skis from oak growing on his ranch. Apocryphal accounts have it that they were of such green wood that they weighed twenty-five pounds. Thompson also cut a long balancing pole, then journeyed to Placerville where he asked the postmaster if there were letters to be delivered in Carson. From that first trip in January 1856 to his last in 1869, he made from thirty to thirty-five Sierra crossings each winter. The eastbound trip usually took three days; the return took two. Aside from letters his pack held supplies for miners at camps along the way—wool socks, tobacco, books, tools, kitchen pans, even seeds ordered from supply houses as spring thaw drew near. Reportedly his loads weighed sixty to eighty pounds, his own part of it little more than a bedroll and a few ounces of dried jerky. In addition to miscellaneous delivery services, Snowshoe Thompson also performed rescues. Early in his second year as mail carrier, the postmaster at Strawberry Station, out from Placerville, requested a search for three missing prospectors. Thompson followed the men's boot prints and found them floundering to their chests in deep snow. He brought them in riding one at a time on the tails of his skis.

On his regular mail trip a few days later Thompson passed a lone cabin a few hours from his eastern destination at

Genoa, and hearing moans he dug free the door and stepped inside. There he found a trapper named John Sisson, who had been marooned with almost no food for twelve days. His legs were purple with frostbite. Thompson skied on to get help, returned with volunteers wearing web snowshoes, then went with them back to Genoa, pulling the poor trapper on a hastily contrived sled. The town doctor said amputation was unavoidable, but he had no morphine. Without stopping to rest, Thompson skied westward over the mountains to get it, and returned again to Genoa. Reporting this round-trip mercy trek, the *Nevada Territorial Enterprise* commented that Snowshoe Thompson "did nothing by halves, but hurtled the Sierra."

Thompson never was paid properly for any of his services. Rescues of course were not part of his contract or subject to payment, but postal officials had agreed to pay $750 per season for his routine ski services. Yet the only money actually paid was $80.22 given Thompson in 1859, ten years before he quit making the runs. Why no further payment was made isn't completely clear. Postmasters apparently passed the responsibility back and forth and nobody took action. In 1872 Thompson traveled to Washington, D.C., armed with a petition passed by the Nevada legislature and signed by the governor requesting Congress to award him $600 in back wages. But nothing came of it. Thompson died four years later and is buried at Genoa beneath a stone with a pair of crossed skis on it and the words, "Native of Norway, departed this life May 15, 1876. Gone but not forgotten."

By the 1900s organized sport skiing began to build popularity in America. A National Ski Tournament was held at Ishpeming, Michigan, in 1905, replete with a Suicide Hill jump 393 feet high. By the late 1920s clubs in New York City and Boston had persuaded the New York, New Haven, and Hartford Railroad to run special night ski trains to New England slopes, and a ski school, the first in America, had opened at Franconia, New Hampshire. The daughter of an innkeeper there married an Austrian ski instructor, Sig Buchmayr, and he agreed to teach skiing at the inn as a stimulus for winter business. In 1932 ski jumping and cross-country competition figured in the winter Olympic Games held at

Lake Placid with Franklin Delano Roosevelt present for the opening ceremonies.

The Games added greatly to public interest in skiing, partly because in attendance there was Lowell Thomas, the renowned broadcaster. At Lake Placid he met Erling Strom, a young Norwegian who had come to the United States from a position in the King's Guard, which had included teaching skiing to the royal family. Thomas signed on for lessons with him, and became addicted to the sport. He had grown up high in the Rocky Mountains, accustomed to snow; had reported on the ski troops in Italy during World War I; and in the late 1920s had entertained Prince William of Sweden and skied a little with one of his aides. The lessons from Strom stimulated a preexisting interest, and Thomas' enthusiasm, in turn, spread across the continent. He was under contract to make nightly news broadcasts from New York City but he began taking his wife, a secretary, a radio engineer, and a telegraph operator from ski slope to ski slope, broadcasting from whatever make-do studio he could contrive and paying the wire charges to New York himself.

This indulging of personal pleasure along with professional calling sent word of skiing into households throughout North America, for Lowell Thomas skied his way from the Canadian Laurentians to the California Sierra winter after winter. His reports brought a bonanza of publicity to ski resorts, most of which were just getting underway in the 1930s. For example, the Union Pacific Railway Company had scouted Mount Rainier, Mount Hood, Yosemite, Reno, Jackson Hole, and supposed meccas in Colorado and Utah, then decided on Ketchum, Idaho, as location for a ski development. They christened an outlying stretch of countryside Sun Valley and hit on a new way to move skiers from valley floor to the tops of slopes: a chair lift. Rope tows recently had been pioneered at Woodstock, Vermont, and the only other lifts in use were gondolas suspended from moving cables. There was nothing between the two extremes. An engineer with the railway, who previously had worked in the tropics loading bananas, got the idea of adapting the type of endless cables he used there into a ski lift. All that was needed was to replace the banana hooks with hanging chairs. Sun Valley opened in 1936, its chair lift the first in the United States—although the California gold-miners-turned-skiers also used to ride seated to the head of one particular run. The mine at Eureka Peak was situated high above its stamp mill and connected by a heavy cable strung with ore buckets. Skiers appropriated the tram for their own purposes. They climbed into empty buckets and rode to the mine tunnel, then jumped out, strapped on skis, and swooped down to the mill for another ride back up.

The adventurous, somewhat hazardous character of downhill skiing has been notorious from the outset. The names of ski runs tell the story: Suicide Six, Nose Dive, Devil's Hangover, Devil's Dip. One of the major forces for safety in the midst of such derring-do, where wearing a leg cast is akin to a badge of honor, is the National Ski Patrol Service. It was born largely of one man's injury and another man's death. Minot Dole, one of the true fathers of modern North American skiing, fell while skiing in the rain at Stowe, Vermont, in 1936. His ankle was badly broken, and he had to lie in the snow for hours while one friend tried futilely to keep him warm and two others went for help. A few weeks later, while Dole still was hobbling around on crutches, one of the friends who had helped rescue him was killed in an interclub ski meet. Grieved and shocked, Dole and others in his ski club decided to study the causes of accidents, then find a way to lessen them.

Out of this grew the present-day National Ski Patrol. An attempt at such a system had been made before, but it operated ineffectively. Dole determined to have a model system of safety and rescue ready by the time of the 1938 downhill and slalom ski races to be held at Stowe. Competitors and spectators were coming from all over America and Europe, and he planned to impress them and set a precedent for future races. Instead of patrolmen simply "being around" and helping if they saw an accident, the new organization put separate teams in charge of definite sections of trail and equipped each with its own toboggan, splints, bandages, blankets, and vacuum bottles of hot coffee. The result is the highly respected volunteer Ski Patrol, which assures reasonable safety and rescue for all.

Minot Dole also was responsible for launching a second organization singularly important in the annals of American skiing. This is the U.S. Army's 10th Mountain Division. In a way it stemmed from the Ski Patrol. In 1939 patrolmen were enjoying beer after races at Manchester, Vermont, and their talk turned to the Finnish ski soldiers who that winter were dismaying the Russian invaders of the Karelian Isthmus. What if foreign troops were to attack the east coast of the United States? Shouldn't there be American ski troops trained for such eventuality? The National Ski Association offered the idea to the War Department, which answered politely with "thanks for your patriotic suggestion" and did nothing.

Dole wasn't to be put off by this brushoff, however. In June 1940 he got the permission of all units within the National Ski Patrol to offer their services to the War Department. The Ski Patrol of Great Britain had been welcomed by the British army, and patrols from ski clubs in Scotland were carrying rations and supplies to antiaircraft stations when mechanical carriers stood snowbound and helpless. Dole knew his idea was good, but it took two sets of contacts through former Yale classmates before he managed to reach General George Marshall and successfully present his proposal. Military thinking at the time held that if Germany were to attack the United States, the most likely route would be down the St. Lawrence River and into the Champlain Valley, as the British had done in 1779. If such an attack were to be stopped the Adirondack, Green, and White mountains would become the line of defense, and in winter that meant ski troops. Skiers native to New York, Vermont, and New Hampshire who knew the overall terrain and the back roads could be organized into patrols and used as scouts and guides.

By April 1941 the Winter Warfare Board approved specifics for ski-troop equipment and training, and search for a suitable army base got underway. Yellowstone National Park, Wyoming, seemed promising. It had been administered by the Army before the National Park Service was established in 1916; it had snow, a certain remote accessibility, and facilities for housing and feeding troops. It also had swans.

Because of overhunting, trumpeter swans were struggling for survival as a species and the hot-spring-heated waters of Yellowstone and the marshes of nearby Red Rock Lakes were their last remaining wintering and breeding grounds. Army maneuvers threatened to vanquish the peaceful white birds forever, and rather than risk that the Army surrendered its preference for Yellowstone.

Fort Lewis near Tacoma, Washington, was second choice. The 87th Mountain Regiment had been set up on paper, and letters to the Ski Patrol requested that as troops were needed, the Patrol supply them. Mount Rainier is only fifty miles south of Fort Lewis, and its Silver Skis tournaments at five-thousand-foot Paradise Valley already were known among skiers. The Army's most pleased recruits surely must have been those ordered to Paradise to ski. By day they donned the skis and clothing newly developed in the Army's winter test program, and by night they perpetuated the camaraderie that always has been part of skiing.

The following winter the United States declared war, and ski training moved to a special camp set up for the purpose in Colorado: Camp Hale. The rest is history. Men there skied hard and trained hard, and when the volunteer 10th Mountain Division was sent to Italy they fought hard. Three months of combat brought four thousand casualties among seven thousand men. Press coverage of the valiant troops for the first time drew the entire nation's attention to skis.

After the war the exuberance of 10th Mountain Division men together with their heroes' stature added impetus to the growth of skiing, which had reached its time for mass popularity. Some of the men returned to lives as ski instructors, others as five-days-a-week working men and two-days-a-week skiers, or ski bums. All acted as catalysts. Aspen had been discovered during Camp Hale days, 10th Mountain Division men favoring its undeveloped powder for off-duty skiing despite difficulties in getting there. As a ski resort, Aspen opened on January 8, 1947, with a parade that featured an honor guard carrying the 10th Mountain Division colors ahead of a horse-drawn sleigh with two colonels and a general. Runners sparked against every cobblestone of the route because there was no snow. It didn't seem a good omen, but it has proven so.

Squaw Valley, California, opened two years after Aspen. Ten million dollars in private money, nine million in state money, and five million in federal money have underwritten its success. So has a white winter blanket so deep that locals apologize if too many pine tops peek above the snow, for this is the region of the Donner Party and of Snowshoe Thompson. New Mexico and Arizona now also boast ski resorts. New Orleans and Dallas have ski clubs, if no snow slopes. Even Hawaii rides the ski bandwagon. On the Big Island the white slopes of Mauna Kea float above the green of palms and the blue of the ocean, and enthusiasts forswear wet suits in favor of ski parkas. Jeeps climb to the 13,784-foot summit with skiers, then grind back down to bring them up again after their downhill runs. The snow often lasts into July.

The world's longest groomed slopes are at Savognin in the Swiss Alps, where even a mini run is more than a mile long, and anybody who can stave off exhaustion can get in fifty miles of lift-served downhill skiing in a day. Iran has some of the newest major resorts. Three are poised above Teheran today, and a fourth is underway. Mountains higher than the grandest peaks of the Alps make Iran a natural for such development. So does a shah who as a schoolboy at St. Moritz skied the Engadine Valley as part of the daily curriculum. Or, if the Caspian region holds no allure, skiers now can try the Caucasus. Russia's Citizen Exchange Corps arranges ski trips to Chegut, largest winter resort in Russia, with six chair lifts, runs featuring vertical drops of 3,400 and 4,300 feet, and a lodge at 13,500 feet on Mount Elbrus, Europe's highest peak, elevation 18,481 feet. If all of this isn't enough, skiers can shuttle back and forth between northern and southern hemispheres; July in the Rockies corresponds to January in the Andes, in Australia, or on Mount Kilimanjaro and professional skiers and Olympic aspirants regularly head south when the ski slopes of North America, Europe, and Japan melt.

Equipment has become as elaborate as has getting to the

slopes. Read an advertisement for "high performance ski-wear," and you know that the baggy pants of the 1950s and the stretch pants of the 1960s have no place in the 1970s or 1980s. Today's fashion calls for "Racy, high-waisted suspendered pants with stretch inserts to give ski-flex." Also fashionable are "Feather Weather parkas of down. The color: pale Olympink for her, darker burgundy Skianti for him. To keep them both warm: nylon plus mylar polyester interlining in the warmups [overpants] and parkas. To keep them dry: rain and stain repellent."

Leather boots are scarcely even a memory. Plastic boots have replaced them on the slopes, the first ones so miserable to wear that foam was custom injected into them in an effort at individualizing fit. It didn't work. Now the attention is on flow padding. This is a layer of silicon with particles, such as cork, floating in suspension. Heat and pressure from the foot cause the silicon to flow and presumably to accommodate to the individual foot. Quiltlike pockets keep it from flowing too far. The boots themselves are two-piece shells of rigid plastic. One piece encases the foot, the other circles the leg as a collar, and a hinge at the heel joins the two. These aren't meant to walk in. They hold the foot to the ski and give optimum control, and that is their whole purpose. One "rear-entry model that is leather-lined and has an adjustable forward lean" retails for $250. An "economical beginners' boot" is offered at $50. One manufacturer has built a spring into the bootheel to "cushion the slam-bam of racing down an icy course." Another boot features flow fit "at the crucial heel area with further adjustment available simply by turning the dial of the contoured instep plate."

Skis themselves also have changed. Their evolution has been from wood to metal to fiber glass to laminations of all three. Metal gives strength for quick, high-powered turns. Fiber glass cuts down on vibrations and clatter and damps excess springiness. Both are lighter than wood, but wood still is used for skis in combination with the newer materials. Moderate-priced skis often have wood cores and glass wraps. One model marketed as CVL, which stands for "constrained viscoelastic layering," consists of three layers of glass with a wooden core, two layers of polyethylene, and an encasing band of steel. These sell at around $200. Another combination has the following layers top to bottom: plastic, steel, aluminum, particle board, aluminum, steel, plastic. Bond these with phenolic resin and mold under 250 pounds pressure per square inch for one hour at 325° F. The steel gives strength. The particle board dampens resiliency. The layers of aluminum determine stiffness, depending on how thick they are. And the plastic makes a durable outer surface with good running qualities.

Even ski slopes are now the product of technology. Wholly artificial slopes offer year-round skiing. Los Angeles has seven-acre Ski Villa with plastic snow made of interlocking tiles, each six inches square and shaggy with bristles that provide a surface skiers say is smoother than actual snow. Similar indoor slopes in Tokyo give office workers a chance to ski during their lunch hours, and portable versions with "moving carpets" are set up from time to time to treat passing New Yorkers to a demonstration of the latest in skis and ski clothes. Special short skis intended for grass rather than snow are produced in Germany—and they gained sudden popularity during the snowless winter of 1976–77 in western United States. City parks with hills of fifteen degrees or more let skiers enjoy their favorite sport in a form described by one ski shop owner as a cross between roller skating and downhill snow skiing. In West Germany itself a permanently white mountain rises above the green fields and forest of Hirschau, the year-round destination of skiers. The great mound is quartz sand, waste product of a large china factory. It was generally considered an eyesore until somebody discovered that the grains of sand have the same skiing properties as snow, except for not melting. A lift like an odd overgrown rowboat now drags skiers to the tops of runs and water sprays onto the slopes at several points to give the quartz added slipperiness and to keep down dust.

Not even real snow is free of the artificial touch. For example in Michigan, which draws about 10 percent of all U.S. skiers, most winter resorts add artificial snow to whatever the clouds deliver naturally. Special machines spew a fine spray of water that freezes into an acceptable facsimile of snow. For many resorts the machines are the only assurance of snow as early as Thanksgiving and as late as Easter, and that long season is needed for profit (which amounts to about $200 million for the entire Michigan ski industry, including equipment and clothing manufacturers and retailers, restaurateurs and gasoline station owners, as well as resort operators). To cover one acre with an inch of such snow requires 27,500 gallons of water, and some ski slopes receive from forty-eight to sixty inches of the artificial snow during the course of a single winter season, although half that is more usual. Underground wells provide the water.

Even where man-made snow isn't added, ski slopes are groomed intensively. Scientific understanding of snow metamorphism elevates the process to an art. A case in point is the Olympic Winter Games held at Sapporo, Japan, in 1972. The structure, crystallography, and density of snow for the race courses all were planned in advance, then brought about by having men rework the snow by foot-packing, ski-packing, and machine-packing. International-caliber racing is highly erosive, and if a course isn't to wear away it must be just about the hardness that would prevent a man from driving a hand shovel into snow. The steepest parts of slalom courses need even greater hardness, so at Sapporo water was mixed into the snow and allowed to freeze.

The tailoring had to begin with the first autumn snowfall, to be sure of the needed strength throughout the entire snow layer, planned for one foot in depth. Since it takes ten to fifteen days for snow to age properly following packing, all that fell closer than that to opening day would have to be removed from the race courses. The government assigned 3,641 men from the Japan Ground Self-Defense Force to Mounts Teine and Eniwa to help get the work done. It was a fortunate decision, because early winter stayed mild and brought little snow to the intended downhill ski courses. About a foot fell in December and the men foot-packed it, then watched a few days later as no new snow fell, and wind started to drift what they had packed. To stop the drifting they cut snow blocks and arranged them over the courses in two-meter grids, the walls standing a meter high at the bottoms of the slopes and a third that at the tops. A skiff of snow fell at New Year's, then was followed by rain that melted part of what previously had been packed to prevent it from drifting.

Still not to be thwarted, the troops deployed into the surrounding forest as far as Eniwa's summit to mine snow and send it down plastic tubes and corrugated aluminum chutes to the race courses. There other men spread the incoming snow, foot-packed it into place, then machine-packed it, and finally side-slipped the slope on skis to smooth the surface. By now January was ending, and opening day was set for February 15. Two small snow-falls came during the nights of January 29–30 and February 2–3, but they were too late to harden by the first race, so the men set to work removing this new snow with shovels and hand plows. Clearing just one downhill course took 250 Self Defense Force men and 100 additional volunteers nearly three hours. In addition to this course, there was another downhill course on Mount Eniwa, four slalom courses on Mount Teine, and cross-country, jump, biathlon, bobsleigh, and luge courses with similar needs. A staggering total of 210,529 man-hours went into three months of preparing and maintaining the courses for all events. More than one quarter million cubic meters of snow were conditioned or removed to bring six and one-half million square meters of snow surface to specification.

The goal was met, and the Olympic committee settled back to enjoy the pleasure their rigid standards and incredible diligence would give the contestants. But it was not to be. Downhill and slalom racers admired the work that had gone in and marveled how beautifully the courses withstood the pressure of the racing, but they didn't like the feel of the snow underfoot. They weren't accustomed to skiing on snow. Race surfaces now usually are prepared by spraying water onto courses and letting it freeze because icy surfaces are fast to achieve and reasonably certain to hold up during a race. Snow packed conventionally and not allowed adequate time to set up gets rutted quickly during competition, and preventing this by grooming snow properly is a great deal of work. Consequently, the world's top racers aren't used to snow. They expect ice, which differs from snow as a high-speed running surface and requires different ski techniques, no matter how perfectly hardened and durable. Sapporo contestants had to adjust to the unexpected snow conditions and that posed problems with nerves, pride, and athletic prowess all at fever pitch.

Daisuke Kuroiwa of the Institute of Low Temperature Science at Hokkaido University and Edward LaChapelle of the Department of Atmospheric Sciences at University of Washington, who worked together on the Sapporo courses, summed up their art thus: "Snow in its original state as it falls from the sky has become a distinct nuisance to these international events—it has to be plowed from highways, prevented from avalanching, shoveled out of the way of spectators and more often than not removed from race courses themselves. . . . From the technical standpoint, it would be easier to process artificial snow generated by snow-making machines into a competition surface than to work with the natural product, for artificial snow is deposited at higher densities than natural snow and crystallographically is already in an advanced state of metamorphism. There seems to be only this one more technological step left before natural snow becomes irrelevant for the Winter Games except for reasons of atmosphere and tradition."

Race speeds now have reached 100 miles an hour and over. Steve McKinney, an American, in 1974 set a record 113.7 mph at Cervinia, Italy, competing against racers from Austria, Canada, Czechoslovakia, Finland, France, Germany, Italy, Japan, Switzerland, and Uganda. At its steepest that course has a sixty-two-degree slope. It runs across three crevasses which are boarded over but nonetheless fling contestants into the air. The course at Portillo, Chile, has a short stretch at eighty degrees. Racers usually hurtle these courses in an aerodynamically efficient position with heads between knees. They can't see and more than one has catapulted off course and been killed. Skintight speed suits, helmets, goggles, and boot covers minimizing the drag of the buckles make up the costume of the racers.

In 1972 with the film *Infinity of Crystals* underway Rick Sylvester skied off El Capitan in Yosemite, one of the high points along the rim. He wore a parachute, and the 1,500-

foot freefall to the bottom of the cliff gave him time to get it open and to take off his skis and drop them. He landed safely, although in pines instead of the intended meadow. Two years before that Yuichiro Miura with $3 million of backing and a retinue of thirty-one comrades and eight hundred porters skied for 6,600 feet and tumbled for 1,320 feet down the South Col of Mount Everest, beginning less than 2,000 feet below the 27,890-foot summit. He too wore a parachute, in fact two of them. The first chute, a small one, popped open when Miura was traveling at 111.8 miles an hour, six seconds after starting. This chute triggered a much larger drag chute which offered his only hope of controlling speed and ultimately stopping. A cross wind spilled the air from it, however, and its effect became that of a whiplashing kite string instead of a brake. Then the wind stopped, and Miura managed to dig in a heel and arrest his fall. He had skied Everest for 98 seconds and skidded, fallen, for another 142 seconds. Eight cameramen recorded the event on 350,000 feet of film, photographing the eighteen-day trek to reach Miura's chosen ski course as well as his climaxing moments.

Why? When the human spirit realizes something might be possible the urge to try grows overwhelming. And it still is one frail man who reaches out when he attains his Everest regardless of the entourage needed to get there, or even the oxygen masks and helmet radios, the financing from equipment manufacturers, and the film documentation for later release to the public. Snowshoe Thompson stretched himself as a person. So did Miura. The difference is in the techniques of the stretching. They have grown complicated. Nonetheless, technology can't ski by itself; that still takes a man.

The blind ski today, encouraged and instructed through a volunteer organization called BOLD, for Blind Outdoor Leisure Development. The program is the idea of Jean Eymers, a former Aspen ski instructor who became blind. Lions Clubs donate special vests to identify blind skiers on the slopes; individual training is based on following the voice of an instructor downhill; and a buddy system takes over after a blind skier has learned the basics. Leg amputees also

ski, aided by special outrigger poles with short ski-like runners. The National Inconvenienced Sportsmen's Association
arranges ski schools and sponsors races.

More than ten thousand skiers now enter Sweden's fifty-
four-mile cross-country Vasalopp, the world's largest ski race.
It was begun in 1921 to commemorate Gustav Vasa who in
1521 skied to the town of Mora to rouse the populace into
rebellion against Danish rule. Winning times for the race
average around five hours. Hot blueberry soup dispensaries
along the course succor contestants willing to pause for
refreshment. Norway has a race from Lillehammer to Rena,
forty miles, commemorating the rescue of a Viking princeling
in 1206. In North America winter resorts from Sugarbush,
Vermont, to Alberta's Banff and California's Yosemite have
marked tracks for cross-country skiing and have hosted races.
Compared to Miura's kamikaze race with his own spirit, or
the intensive grooming of Olympic Games snow courses, all
such cross-country skiing seems low key. Equipment is less
specialized than for racing, clothing is less controlled by
fashion, the skiing itself is whatever you care to make it—
gentle or vigorous, solitary or competitive.

One early spring not long ago a former ranger associate
of ours named Wayne Merry stopped to visit. He was en
route home to Yosemite where he headed the Mountaineering Guide Service. He was coming from Alaska, having just
finished skiing with three friends from Bettles through Gates
of the Arctic to the Beaufort Sea, a 300-mile winding course
that took the men 180 straightline miles. Wayne spoke
quietly of the trip. Snow travel has advantages: Water is
always available if you take time to melt it. And you're
guaranteed a level campsite: By scraping you can tailor
practically anyplace to suit your needs. Of course there are
disadvantages. Snow can be so soft you sink to your knees,
and it takes till the third man in line before the trail is
adequately packed for skiing instead of thrashing around.
Or sometimes your skis break through powder to depth hoar
and again you're floundering, with no possibility of waxing or
otherwise adjusting for such a surface. Yet what an experience. Take just the sounds. There's the creak of your gear,
the squeak of the snow, the play of the wind. At night

there's even the rhythmic thunder of your whiskers hitting with every breath against the nylon cover of your sleeping bag. And occasionally there comes the chorusing of wolves, harmonics like organ chords built into their voices, night music of the frozen north since time began.

Talking with Wayne, I decided to join the last Snow Survival course of the season to be given at Yosemite. Wayne's assistant Mead Hargus led the weekend course I signed into. He began with a laconic, "Well, we're going out to dig some snow caves and learn all about hypothermia and things like that." Then he passed around mimeographed sheets on "Rules for Avoiding and Surviving Avalanche" and a chart listing standard distress signals. "These are great if somebody is looking for you," Mead commented, "but it's better to rely on yourself."

By afternoon nine of us had skied a few miles beyond Badger Pass and begun to dig our caves for the night. At one point I remember thinking I'd probably always stick with a tent, given a choice. Snow caves aren't difficult to dig but they are slow. They are worth the bother in an emergency or for prolonged camping in one place, but too time-demanding for one-night camps. You think this mostly at first, struggling to dig without enough space for working. A snow cave starts with sinking a well into the snow, then tunneling horizontally to hollow a sleeping niche with a smoothly domed ceiling, to prevent dripping. A shelf notched into the cave wall a foot or two above the floor protects you from the downward drainage of cold air and assures a comfortable night, given a warm-enough sleeping bag that has been successfully kept dry and is well insulated underneath. Novices sometimes forget the insulation, but it is necessary to prevent losing your body heat to the snow. A pack or a movable snow block set in the doorway will stop wind, and you must slope the entry away from the cave, not down into it. This is to let cold air spill out rather than in. Add a vent hole through the roof and you are done, although it's wise to keep a ski pole stuck through this hole for occasional jiggling at night to break through any new snow that has fallen.

For a while during cave construction, happiness seems

like cutting a big, firm snow block because that's the easiest
way to move many cubic inches of snow at a time. Lacking
proper blocks, which are possible only in certain snow conditions, you seem to shovel snow out the door and up your well
forever. Happiness also is noticing the deep blue of the snow
as you burrow into it, and hearing the utter silence. This
first cave of mine took four hours of work. Experience can
cut that time in half—and, oddly, one snow cave leads to
another. They are habit forming. You leave the warmth of
campfire and companionship and crunch through the snow
to where you remember your hole awaits. The pines and
incense cedars stand as giant black plumes against white
slopes and starry sky. You wriggle into your nest, taking all
gear with you except skis which are standing upright in
the snow, too tall to be covered by any new-fallen nighttime
blanket. Then you slip into your sleeping bag fluffed out on
the bench, which is contoured to fit hips and shoulders. A
single candle gives ample light if you care to read. The
porous nature of the snow and your vent assure fresh air.
You sleep sealed off from outside sounds and cold.

I remember lying there that first night thinking of ptarmigan also sleeping within the snow, and of ages-old Eskimo snow igloos lived in for months at a time and enlivened
with the births of babies and the drumbeats of dances.
Thoughts came to mind of Sir Charles Wright navigating
with Scott across the antarctic barrens by feeling the sastrugi
with his feet and using only one eye at a time, keeping the
other shut and in reserve in case of injury. I thought of
dogsled races in Alaska today, and of North Dakota blizzards, and of skiing to remote patrol cabins at Mount Rainier
to cut snow from the roofs; also about proliferating snowmobiles and ski developments that impact village traffic and
require elaborate sewage drain fields and disposal plants.
Snow now fosters complicated human actions and aspira-
tions. Snow also still cradles simple responses and serene
awareness of winter's purity.

Sleep within a Yosemite snow cave, and you find peace.

ACKNOWLEDGMENTS

IF, AS I BELIEVE TO BE THE CASE, NO NONFICTION BOOK IS truly a product of the person who writes it, the situation applies doubly and triply in this instance. Indebtedness for this book extends to the specialists whose reports I have read and to those who have taken time to talk with me or to criticize drafts of chapters.

Sir Charles Wright, next-to-last survivor of Scott's second antarctic expedition, allowed me to visit him at his island home in Canada shortly before his death, and his artist daughter, Pat Wright, has helped with manuscript details in the months since then. Dr. Edward LaChapelle, professor of atmospheric sciences at the University of Washington, I first met in a tunnel dug to bedrock beneath the icefall of the Blue Glacier on Mount Olympus. There, with various gauges, he was watching both the glacier's movement downslope and its pressure slowly squeezing the tunnel shut. Ed is expert on the subjects of avalanches and of snow crystals and metamorphism, as well as glaciology, and he has been endlessly helpful with this book. Yorke Edwards, a mammalogist who now directs the British Columbia Provincial Museum, sent for certain reports I couldn't obtain (particularly those of Russian snow-ecologists), stimulated my thinking, and read manuscript for me. Dr. Robert Ackerman, anthropologist at Washington State University with a specialty in arctic archaeology, and Dr. Wallace Cady of the U.S. Geological Survey, also carefully read and commented on chapters, as did Glenn Gallison of the National Park Service.

Jim Whittaker, famed conqueror of Everest and personal acquaintance from mutual long-ago days at Mount Rainier, took time out from preparing for a new Himalayan assault to talk about snow and clothing; and Wayne Merry, former chief ranger at Mount McKinley National Park and head of the mountaineering guide service in Yosemite, not only recounted experiences but invited me on a ski trek. On trips to Japan various people contributed bits of information, and Yoshi Nishihara who lives with my husband and me in Tacoma and her mother, Kei Nishihara of Osaka, gathered written material and translated it. In the Far North the Wik, Jones, and Denslow families introduced us to arctic life, including snowmobiling and dog-sledding; and Ole Wik in particular has read and reread draft chapters. Eskimo people in villages along the Kobuk River bestowed patience and cordiality, and Bob Uhl of Kotzebue generously gave of his time, counsel, and firsthand information concerning arctic snow conditions. Similarly, Gunther Lishy, a fur trapper along the Yukon–British Columbia border, shared his knowledge and experience. Chess Lyons, formerly forest engineer and naturalist with the British Columbia provincial government, read manuscript and buoyed morale.

Listing these persons is pleasant for it stirs memories of shared experiences and conversations, and because it pleases me to contemplate the human willingness to help. Listing also is worrisome, however, lest I have failed to properly convey information and understanding entrusted to me. Inaccuracies or inadequacies within the book are of course my own doing. To those who helped me think, and research, and—at least to some extent—know, I am grateful in the extreme. My fervent hope is that these pages justify their kindness.

Countless companions on excursions into the snowy hinterlands are hereby also acknowledged gratefully and happily, most especially my husband, Louis, whose love of the outdoors, snowy and otherwise, has infected, enriched, and molded my life and thinking.

SELECTED BIBLIOGRAPHY

HE LITERATURE OF SNOW IS SCATTERED THROUGHOUT A SURPRISING NUMBER
: distinct disciplines. Space precludes a complete list of sources consulted for
ɪis book, many of them reports in obscure journals unlikely to be of general
ɪterest. The citations given are intended as a broad sampling of available
ʹferences; they indicate the scope of published material and, through the
tations accompanying most of them, will point toward additional sources.

General References

ᴇʟʟ, Cᴏʀʏᴅᴏɴ. *The Wonder of Snow.* New York: Hill and Wang, 1957.
astern Snow Conferences, proceedings of the meeting published annually by
the host institution, which varies.
ɪɴɢᴇʀʏ, W. D. *Ice and Snow: Properties, Processes, and Applications.* Cambridge: Massachusetts Institute of Technology, 1963.
ʀoceedings of the 1966 Helsinki Symposium on the Ecology of the Subarctic
Regions. UNESCO *Series in Ecology and Conservation,* No. 1, Paris, 1970.
ᴀɴᴛᴇғᴏʀᴅ, Hᴇɴʀʏ S., and Jᴀᴍᴇs L. Sᴍɪᴛʜ, compilers. "Advanced Concepts
and Techniques in the Study of Snow and Ice Resources. A United States
Contribution to the International Hydrological Decade." Washington, D.C.:
National Academy of Sciences, 1974.
ɪnow and Ice in Relation to Wildlife and Recreation, a Symposium." Ames:
Iowa State University, February 11-12, 1971.
ɪnow Hydrology." Canadian National Committee, The International Hydrological Decade. Proceedings of a Workshop Seminar, University of New
Brunswick, 1968.
/estern Snow Conferences, proceedings of the meeting published annually
by the host institution, which varies.

Climate

ᴏuɢʟᴀs, Jᴏʜɴ H. "Climate Change: Chilling Possibilities," *Science News,*
107 (1975): 138-40.
ᴀᴛᴇs, W. R. et al. "Variations in the Earth's Orbit: Pacemaker of the Ice
Ages." *Science* 194, 4270 (1976): 1121-32.
ᴇɪɢᴇʀ, Rᴜᴅᴏʟғ. *Climate Near the Ground.* Translated by Scripta Technica,
Inc. Cambridge: Harvard University Press, 1965.
ᴀᴀɢ, Wɪʟʟɪᴀᴍ G. "The Bering Strait Land Bridge." *Scientific American,*
January 1962, pp. 112-20.
ᴏʟᴍᴇs, Aʀᴛʜuʀ. *Principles of Physical Geology.* New York: The Ronald
Press, 1965.
ᴏᴘᴋɪɴs, Dᴀᴠɪᴅ Mᴏᴏᴅʏ. *The Bering Land Bridge.* Stanford: Stanford University Press, 1967.
ᴜᴋʟᴀ, Gᴇᴏʀɢᴇ J., and Hᴇʟᴇɴᴀ J. Kᴜᴋʟᴀ. "Increased Surface Albedo in the
Northern Hemisphere." *Science* 183, 4126 (1974): 709-14.
ᴜʀᴛᴇ́ɴ, Bᴊᴏ̈ʀɴ. "Pleistocene Mammals and the Bering Bridge." *Commenta-
tiones Biologicae.* 29,8 (1966): 35-41. Societas Scientiarum Fennica.
———. *Istiden.* Stockholm: International Book Production, 1969.
ʜuᴍsᴋɪɪ, P. A. *Principles of Structural Glaciology,* translated from Russian by
David Kraus. New York: Dover Publications, 1964.

WALLACE, ROBERT. "A Viking Village in America." New York: Time-Life Pu' lishers. *Nature and Science Annual,* pp. 16-31, 1975.

Polar Regions and Far North

CHERRY-GARRARD, APSLEY. *The Worst Journey in the World.* London: Chat* and Windus, 1922, reprinted 1952.
DEBENHAM, FRANK. *Antarctica.* New York: Macmillan, 1961.
KIRMAN, L. P. *The White Road.* London: Hollis and Carter, 1959.
LLANO, GEORGE A. "The Terrestrial Life of the Antarctic." *Scientific America* September 1962, pp. 212-18.
MOWAT, FARLEY. *Ordeal by Ice.* Boston: Little, Brown and Co., 1960.
NANSEN, FRIDTJOF. *Through Siberia, the Land of the Future.* London: Willia: Heinemann, 1914.
STONEHOUSE, BERNARD. *Animals of the Arctic: the Ecology of the Far Nort* New York: Holt, Rinehart, and Winston, 1972.

Snow Crystals and Avalanches

ATWATER, MONTGOMERY M. *The Avalanche Hunters.* Philadelphia: Macra Smith, 1968.
BENTLEY, WILSON ALWYN, and W. J. HUMPHRIES. *Snow Crystals.* New Yor* McGraw-Hill, 1931. Reprinted by Dover Publications, New York, 1964.
FRASER, COLIN. *The Avalanche Enigma.* New York: Rand McNally, 1966.
FRIEDL, JOHN. "Swiss Family Togetherness." *Natural History,* April 1977, p* 40-45.
LACHAPELLE, EDWARD R. *The ABC of Avalanche Safety.* Denver: Colorad Outdoor Sports Co., 2nd revised edition, 1970.
————. *Field Guide to Snow Crystals.* Seattle: University of Washington Pres* 1969.
LACHAPELLE, E. R. et al. "Avalanche Studies." Olympia: Washington Stat Highway Department Research Program, Report 8.4. 3rd Annual Repor November 1973.
MAYKUT, GARY. "Snow Seminar." Water Resources Research Institute. Co: vallis: Oregon State University, 1969.
MELLOR, MALCOLM. "Avalanches." Cold Regions Science and Engineerin* Part 3, Snow Technology. Hanover, New Hampshire: Cold Regions Researc and Engineering Laboratory, 1968.
NAKAYA, UKICHIRO. *Snow Crystals, Natural and Artificial.* Cambridge: Harvar University Press, 1954.
SELIGMAN, GERALD. *Snow Structure and Ski Fields.* London: Macmillan an* Co., 1936.

Water

FIELD, WILLIAM O. "Glaciers." *Scientific American,* September 1955, pp. 84-9*
MEIR, MARK F. "Glaciers and Water Supply." *Journal of American Wate Works Association,* 61, 1 (1969): 20-29.
WORK, R. A. et al. "Accuracy of Field Snow Surveys, Western U. S. includin* Alaska." Technical Report 163. Hanover, New Hampshire: Cold Region Research and Engineering Laboratory, August 1965.

Snow Removal

CHURCH, J. E. "The Human Side of Snow." *Science Monthly* 54 (1942): 211-2*
Department of Sanitation, miscellaneous publications. Public Informatio Office, New York.
"Environmental Impact of Highway Deicing." Edison Water Quality Lab Edison, New Jersey. Interim report, Environmental Protection Agency Water Quality Office 11040, QCG 1970.

ING, ERNEST L. as told to Robert E. Mahaffay. *Main Line: Fifty Years of Railroading with the Southern Pacific.* Garden City, N.Y.: Doubleday and Co., 1948.

INSK, L. D. "Survey of Snow and Ice Removal Techniques." Technical Report 128. Hanover, New Hampshire: Cold Regions Research and Engineering Laboratory, 1964.

ISBET, I. C. T. "Has Salt Lost Favor?" Lincoln, Mass.: *Conservation Leader,* 1972.

———. "Salty Words on Salty Roads." Massachusetts Audubon *Newsletter* 14, 6 (1975): 8-9.

ttawa Conferences on Snow Removal. National Research Council, Technical Report 83, Ottawa, 1964.

Roads in Winter." *American City* (Town and Country edition) 19 (1918): 102-3.

CHNEIDER, T. R. "Snowdrifts and Winter Ice on Roads." National Research Council of Canada, Technical Translation 1038, Ottawa, 1959.

eminar WR 011.69, Spring Quarter 1969. Water Resources Research Institute. Corvallis: Oregon State University, 1970.

Water Pollution and Associated Effects from Street Salting." Environmental Protection Technology Series. Environmental Protection Agency R2-73-257. Washington, D.C., May 1973.

History

ERTON, PIERRE. *Klondike.* Toronto: McClelland and Stewart, 1972.

NGLE, ELOISE, and LAUIR PAAMANEN. *The Winter War, the Russo-Finnish Conflict, 1939-40.* New York: Charles Scribner's, 1973.

Shelter and Clothing

LSNER, ROBERT W., and WILLIAM O. PRUITT, JR. "Some Structural and Thermal Characteristics of Snow Shelters." *Arctic* 12,1 (1959): 20-27.

OLD, L. W., compiler. Annotated Bibliography on Snow Drifting and Its Control. Division of Building Research. Ottawa: National Research Council of Canada, 1968.

ATT, GUDMUND. "Arctic Skin Clothing in Eurasia and America, An Ethnographic Study." *Arctic Anthropologist* 5,2 (1969): 3-132.

OPPES, WAYNE F. "A Report on Characteristics of Snow Houses and Their Practicability as a Form of Temporary Shelter." Committee on Sanitary Engineering and Environment. Ottawa: National Research Council of Canada, 1948.

WINSON, G. K. "Preliminary Investigations of Permacrete." U. S. Army Material Command, Hanover, New Hampshire: Cold Regions Research and Engineering Laboratory, 1965.

Native Peoples

ALIKII, ASEN. *The Netsilik Eskimos.* Garden City, N.Y.: Natural History Press, 1970.

RIDGES, LUCAS E. *Uttermost Part of the Earth.* New York: E. P. Dutton, 1950.

ROWN, MALCOLM G. "Cold Acclimatization in Eskimos." *Arctic* 7, 3 & 4 (1954): 343-53.

OON, CARLETON S. "Man Against the Cold." *Natural History,* August 1970, pp. 40-47.

OLGIKH, B. O. "Problems in the Ethnography and Physical Anthropology of the Arctic." *Arctic Anthropologist* 3,1 (1965): 1-9.

———. "The Formation of the Modern Peoples of the Soviet North." *Arctic Anthropologist* 9,1 (1972): 17-24.

RVING, LAURENCE. *Arctic Life of Birds and Mammals including Man.* New York: Springer-Verlag, 1972.

BIRKET-SMITH, KAJ. *Eskimos.* New York: Crown Publishers, 1971.

LANTIS, MARGARET. "Problems of Human Ecology in the North America Arctic." *Arctic* 7, 3 & 4 (1956): 307-20.

LEE, RICHARD B., and IRVEN DEVORE. *Man the Hunter.* Chicago: Aldine Publishing Company, 1968.

NELSON, RICHARD. *Hunters of the Northern Ice.* Chicago: University of Chicago Press, 1973.

PELTO, PERTTI, J. *The Snowmobile Revolution: Technology and Social Change in the Arctic.* Menlo Park, Calif.: Cummings Publishing Company, 1973.

SERVICE, ELMAN R. *Profiles in Ethnology.* New York: Harper and Row, 196

STEFANSSON, VILHJALMUR. *My Life with the Eskimos.* New York: Macmilla 1913.

———. *Hunters of the Great North.* New York: Harcourt, 1922.

———. *Arctic Manual.* New York: Macmillan, 1944.

STEWARD, JULIAN H. and LOUIS C. FARON. *Native Peoples of South Americ* New York: McGraw-Hill, 1959.

TAKSEMI, CHUNER. *Nivkh.* Leningrad: Nauka, pp. 194-96, 1967.

VANSTONE, JAMES W. *Point Hope: an Eskimo Village in Transition.* Seattl University of Washington Press, 1962.

WEYER, EDWARD MOFFAT. *The Eskimo: Their Environment and Folkway* New Haven: Yale University Press, 1969.

Sleds, Sled Dogs, and Reindeer Herding

HALL, GUDMUND. "Notes on Reindeer Nomadism." *American Anthropologi* 6 (1917): 75-133.

HONIGMANN, JOHN J. "The Kaska Indians: an Ethnographic Reconstruction Yale University Publication in *Anthropology*, No. 51, 1945. Reprinted b Human Relations Area File Press, 1964.

JACKSON, SHELDON. "Introduction of Domestic Reindeer into Alaska." 13 Annual Report, 1903. Washington, D.C.: Government Printing Office, 190

JENNESS, D. "The Copper Eskimos." Canadian Arctic Expedition, 1913-1 Part 12, Ottawa. 1922.

LANGKAVEL, B. "Dogs and Savages." 18th Annual Report, 1896-97. Washing ton, D.C.: Bureau of American Ethnology, Smithsonian Institution, 189

LANTIS, MARGARET. "The Reindeer Industry of Alaska." *Arctic* 3,1 (1950 27-44.

LAUFER, BERTHOLD. "The Reindeer and Its Domestication." *American Anthr pologist* 4 (1917): 91-147.

MCKENNAN, ROBERT. "The Upper Tanana Indians." Yale University Public tions in *Anthropology*, No. 55, 1959.

NAYLOR, LARRY et al. "Socio-Economic Evaluation of Reindeer Herding i Relation to Proposed National Interest (D-2) Lands in Northwester Alaska." Seattle: U. S. Department of Interior, National Park Service Co tract CX-9000-6-0098, April 1970.

VESEY-FITZPATRICK, BRIAN. *The Domestic Dog.* London: Routledge and Kega Paul, 1957.

WARD, ROBERT H., and DOLLY WARD. *The Complete Samoyed.* New Yor Howell Book House, 1971.

———. *The Complete Siberian Husky.* New York: Howell Book House, 197

ZEUNER, FREDERICK E. *The History of Domesticated Animals.* London: Hutchi son, 1963.

Plant Life

BILLINGS, L. D., and L. C. BLISS. "An Alpine Snowbank Environment and I Effects on Vegetation, Plant Development, and Productivity." *Arctic* 40, (1959): 388-97.

LISS, L. C. "Adaptations of Arctic and Alpine Plants to Environmental Conditions." *Arctic* 15, 2 (1962): 117-44.

ORTLORF, WILLIAM L. D. "Some Forest Influences on Thermal Balance over Snowpack." Army Corps of Engineers. Research Note, Snow Investigations. 1952.

ARMODAY, BARRY BURTON. "Control of Plant Production, Phenecology, and Distribution by a Subalpine Snowbank Microenvironment." Master's Thesis, Bellingham: Western Washington State College, 1973.

ULLER, WILLIAM A., and JOHN C. HOLMES. *The Life of the Far North.* New York: McGraw-Hill, 1972.

OHAM, RONALD WILLIAM. "Laboratory and Field Studies on Snow Algae of the Pacific Northwest." Ph.D. Thesis, Seattle: University of Washington, 1971.

Snow Molds." Washington Agricultural Experiment Station, Bulletin 677. Pullman: Washington State University, 1966.

Bird and Animal Life

ANNIKOV, A. G. et al. "Biology of the Siaga." Moscow, 1961. Translated by Israel Program for Scientific Translations, Jerusalem, 1967.

ATON, GRAY. "Snowball Construction by a Feral Troop of Japanese Macaques (*Macaca fuscata*) Living Under Seminatural Conditions." *Primates* 13,4 (1972): 411-14. Japan Monkey Center, Inuyama, Aichi, Japan.

DWARDS, R. YORKE. "Foods of Caribou in Wells Gray Park, B.C." *The Canadian Field Naturalist* 74,1 (1960): 3-7.

DWARDS, R. YORKE, and RALPH W. RITCEY. "The Migrations of a Moose Herd." *Journal of Mammalogy* 37,4 (1956): 486-94.

———. "Migrations of Caribou in a Mountainous Area in Wells Gray Park, B.C." *The Canadian Field Naturalist* 73,1 (1959): 21-25.

KVALL, ROBERT B. *Fields on the Hoof.* New York: Holt, Rinehart & Winston, 1968.

ORMOZOV, A. N. "Snow Cover as an Integral Factor of the Environment and Its Importance in the Ecology of Mammals and Birds." Translated by W. Prychodko and W. O. Pruitt, Jr. Occasional Paper No. 1, Boreal Institute, Edmonton: University of Alberta, 1966.

ORMOZOV, A. N. "The Ecology of the Most Important Species of Subarctic Fauna. Ecology of Subarctic Regions." UNESCO Ecology and Conservation Series, 1968.

ESSE, RICHARD. *Ecology of Animal Geography.* New York: John Wiley and Sons, 1937.

RVING, LAURENCE. "Adaptations of Cold." *Scientific American*, January 1966, pp. 94-101.

ELSALL, JOHN P. *The Migratory Barren-Ground Caribou of Canada.* Ottawa: Canadian Wildlife Service, 1968.

EVICK, G. MURRAY. *Antarctic Penguins.* London: William Heinemann, 1914.

IECH, DAVID. *The Wolf: The Ecology and Behavior of an Endangered Species.* New York: Natural History Press, 1970.

IITCHELL, JOHN G. "Where Have all the Tuttu Gone?" *Audubon* 79,2 (1977): 2-15.

EDERSON, ALWIN. *Polar Animals.* New York: Taplinger Publishing Company, 1966.

ERRY, RICHARD. *The World of the Polar Bear.* Seattle: University of Washington Press, 1966.

———. *The Polar Worlds.* New York: Taplinger Publishing Company, 1973.

RUITT, WILLIAM O., JR. "Animals in the Snow." *Scientific American*, January 1960, pp. 60-68.

———. "Snow as a Factor in the Wintering Ecology of the Barren Ground Caribou." *Arctic* 12,3 (1959): 159-79.

PRUITT, WILLIAM O., JR. *Animals of the North.* New York: Harper and Row 1960.

SCHOLANDER, P. F. et al. "Adaptations to Cold in Arctic Mammals and Birds Relation to Body Temperature, Insulation, and Basal Metabolism." *Biologic Bulletin* 99,2 (1950): 259-71. Ottawa: Canadian Wildlife Service.

SMITH, FRANK, and PAUL S. WELCH. "Oligochaeta Collected by the Canadia Arctic Expedition 1913-18." Volume 9, Ottawa, 1919.

STOGANOV, S. J. "Carnivorous Mammals of Siberia." Izdatel'stvo Akademii Nat SSSR, Moskva, 1962. Israel Program for Scientific Translations, Jerusalem 1969.

TYMEN, MICHAEL I. "The Geographical Distribution of Ice Worms." *Canadic Journal of Zoology* 48,6 (1970): 1363-67.

SYMINGTON, FRASER. "Tuktu, the Caribou of the Northern Mainland." Ottawa Canadian Wildlife Service, 1965.

Snowmobiles

BALDWIN, MALCOLM F., and DAN H. STODDARD, JR. "The Off-Road Vehic and Environmental Quality." Washington, D.C.: The Conservation Found tion, 1973.

MOLLER, GEORGE H. "The Landowner and the Snowmobiler." U.S. Departmen of Agriculture, Forest Service Research Paper NE-206. Northeastern Fore Experiment Station, Upper Darby, Pennsylvania, 1971.

"Planning Considerations for Winter Sports Development." U. S. Departmen of Agriculture, Forest Service, in cooperation with the National Ski Are Association, 1973.

SMITH, LORNE. "The Mechanical Dog Team: A Study of the Skidoo in th Canadian Arctic." *Arctic Anthropologist* 9,1 1972): 1-9.

"Snowmobile and Off the Road Vehicle Research Symposium." Department Parks and Recreation Resources, Agricultural Experiment Station, U. S. D partment of Agriculture, Bureau of Outdoor Recreation. Michigan Stat University, Technical Report 8, 1971.

WANEK, WALLACE J. "A Continuing Study of the Ecological Impact of Snov mobiling in Northern Minnesota." Research reports for 1971-72, 1972-7 and 1973-74, Center for Environmental Studies, Bemidji State College.

WATSON, J. ALDEN, and MORLAN W. NELSON. "A History of the Developmen of Oversnow Vehicles." Western Snow Conference, 1968.

Skis and Snowshoes

DAVIDSON, D. S. "Snowshoes." Memoirs of the American Philosophical Societ vol. 6, 1937.

DOLE, MINOT. *Adventures in Skiing.* New York: Franklin Watts, 1965.

HUNT, JOHN CLARK. "Where American Skiing Was Born: Would You Believ Rabbit Creek?" *Westways Magazine,* Los Angeles: Auto Club of Southen California, December 1973, pp. 24-27, 70-72.

International Symposia on the Role of Snow and Ice in Hydrology; Measure ment and Forecasting. WMO 5 Modification of Snowfall, Snowcover, an Ice Cover. n.d.

JACKSON, JOSEPH HENRY. *Anybody's Gold: the Story of the California Minin Towns.* New York: D. Appleton Century, 1941.

KUROIWA, D., and EDWARD R. LACHAPELLE. "The Preparation of Artificia Snow and Ice Surfaces for the 11th Olympic Winter Games, Sapporo." n.c

MCKINNEY, STEVE, with DICK DORWORTH. "How I Broke the World Spee Record." *Ski,* Spring 1975, pp. 30-35.

WARD, HUBERT WARREN. "The Sporting Scene on Holmenkollen Hill." *Th New Yorker,* February 28, 1977.

WISBY, JOHANNES HROFF. "Carrying the Mail over the Andes on Skis." *Outin* 37 (1901): 672-5.

INDEX